Multi-Pronged Omics Technologies to Understand COVID-19

Multi-Pronged Omics Technologies to Understand COVID-19

Edited by Sanjeeva Srivastava
Professor, Department of Biosciences and Bioengineering,
Indian Institute of Technology Bombay, India

CRC Press
Taylor & Francis Group
Boca Raton London

CRC Press is an imprint of the
Taylor & Francis Group, an **informa** business

First edition published 2022
by CRC Press
6000 Broken Sound Parkway NW, Suite 300, Boca Raton, FL 33487–2742

and by CRC Press
2 Park Square, Milton Park, Abingdon, Oxon, OX14 4RN

CRC Press is an imprint of Taylor & Francis Group, LLC

Library of Congress Cataloging-in-Publication Data
Names: Srivastava, Sanjeeva, editor.
Title: Multi-pronged omics technologies to understand COVID-19 /
edited by Sanjeeva Srivastava.
Description: First edition | Boca Raton : CRC Press, 2022. |
Includes bibliographical references and index.
Identifiers: LCCN 2021059642 (print) | LCCN 2021059643 (ebook) |
ISBN 9781032116266 (hbk) | ISBN 9781032116297 (pbk) | ISBN 9781003220787 (ebk)
Subjects: MESH: COVID-19—genetics | SARS-CoV-2—genetics | Computational Biology
Classification: LCC RA644.C67 (print) | LCC RA644.C67 (ebook) | NLM WC 506 |
DDC 614.5/92414—dc23/eng/20220314
LC record available at https://lccn.loc.gov/2021059642
LC ebook record available at https://lccn.loc.gov/2021059643

ISBN: 978-1-032-11626-6 (hbk)
ISBN: 978-1-032-11629-7 (pbk)
ISBN: 978-1-003-22078-7 (ebk)

DOI: 10.1201/9781003220787

Typeset in Times
by Apex CoVantage, LLC

Contents

Preface

The World Health Organization (WHO) declared COVID-19 (also referred to as SARS-CoV-2 disease) as a global pandemic on March 11, 2020. Since then, COVID-19 infections have been spreading at an alarming rate globally. The high transmission rates, limited diagnosis, and no effective therapy pose a considerable challenge in controlling and diminishing the pandemic. Scientists, researchers, and healthcare professionals worldwide are racing to use the best possible technologies and therapeutic strategies, which are being expedited in an unprecedented way. Along with this outbreak, 2020–2021 has also witnessed omics, bioinformatics, and machine learning revolution making advances in high-resolution mass spectrometry, next-generation sequencing (NGS), microarray technologies, and other spectroscopy-related methods. The scientific community continues to gather data and share information across the globe through preprint servers as well as peer-reviewed scientific publications. Several federal agencies freely share COVID-19 open access data and other resources with the research community. This book summarizes the recent advances in understanding SARS-CoV-2 disease using the latest omics-based technologies and inventions.

Chapters 1 and 2 focus on the epidemiology and strength of NGS technology for tracking mutations across the globe. These chapters summarize the epidemiological characteristics, clinical features, and detection of emerging mutations of SARS-CoV-2 that have played a crucial role in public health decisions. Chapters 3, 4, and 5 summarize the use of high-throughput mass spectrometry technology for the detection and prognosis of COVID-19 progression. These technologies have been applied for the in-depth proteomics and metabolomics study of the COVID-19-infected host. Such studies will aid in providing valuable insights into the altered proteome and metabolome of the host leading to severe clinical manifestation. Moreover, the panel of potential biomarkers can be used in the clinics for the prognosis of disease outcomes before it could lead to fatal symptoms. Currently, the proteome microarray technique is widely explored as a robust diagnostic platform to detect infectious diseases, including COVID-19. Chapter 6 outlines the recent development in microarray techniques for the detection and monitoring of diseases. Chapters 7 and 8 summarize the understanding of the host immune response toward the infection and the impact of SARS-CoV-2 on different organs of the body post-recovery. These chapters thus discuss the role of multi-omics technologies for the prognosis and management of the long-term impact of COVID-19 infection.

The application of omics technology for understanding the mechanism of COVID-19 infection has revealed several potential targets for therapeutic interventions. Chapter 9 summarizes the *in silico* drug repurposing for combating COVID-19. Chapter 10 provides an overview of several other spectrometry-based methods used to detect the SARS-CoV-2 virus from the clinical samples. With the recent development of machine learning technology and the rate at which COVID-19 omics data accumulates, researchers have also tried to use the freely available resources to build the COVID-19 prediction model. Thus, Chapter 11 summarizes the use of big data and machine learning–based approaches to predict the clinical markers of COVID-19.

This book aims to summarize valuable insights from the recent resources on applications of omics-based technology for the management of COVID-19. These resources lay the foundation for understanding the mechanism of SARS-CoV-2 infection in humans and allow us to examine the constantly evolving field of omics. This book will provide the readers with a broad scientific outlook on the future scope of omics and their application in managing SARS-CoV-2 infections.

Editor biography

Sanjeeva Srivastava is a professor and the group leader of the Proteomics Laboratory at the Department of Biosciences and Bioengineering, Indian Institute of Technology Bombay, India. High-throughput proteomics, protein microarrays, and mass spectrometry are among his specialties. He has implemented ground-breaking AI-driven data analytics on big biological datasets. He has been at the forefront of clinical and biomedical research based on big data. His group's current research focuses on the development of clinical biomarkers for infectious diseases and malignancies. His group has also pioneered therapeutic target identification efforts and decoded protein interaction networks in human illnesses like gliomas and contagious diseases like COVID-19 and malaria. His group has developed diagnostic biomarkers and described the pathophysiology of severe malaria (falciparum and vivax) and COVID-19, especially the underlying mechanisms that lead to the development of severe sequelae.

Dr. Srivastava is an active contributor to global proteomics research and development. He serves on the Executive Committee of Human Proteome Organization (HUPO) and Proteomics Society, India (PSI). He has more than 135 publications from his work as an independent researcher at IIT Bombay. To date, he has filed 19 patents that include biomarkers for various types of cancers, infectious diseases like malaria, leptospirosis, COVID-19, and proteomics method development in uncharted territories such as fish and plant proteomics.

Contributors

Arup Acharjee
Department of Biosciences and Bioengineering
IIT Bombay
Mumbai, Maharashtra, India

Shalini Aggarwal
Department of Biosciences and Bioengineering
IIT Bombay
Mumbai, Maharashtra, India

Graham Roy Ball
School of Science and Technology
Nottingham Trent University
Nottingham, UK

Arghya Banerjee
Department of Biosciences and Bioengineering
IIT Bombay
Mumbai, Maharashtra, India

Nirjhar Banerjee
Department of Biosciences and Bioengineering
IIT Bombay
Mumbai, Maharashtra, India

Renuka Bankar
Department of Biosciences and Bioengineering
IIT Bombay
Mumbai, Maharashtra, India

Abhilash Barpanda
Department of Biosciences and Bioengineering
IIT Bombay
Mumbai, Maharashtra, India

Surbhi Bihani
Department of Biosciences and Bioengineering
IIT Bombay
Mumbai, Maharashtra, India

Deeptarup Biswas
Department of Biosciences and Bioengineering
IIT Bombay
Mumbai, Maharashtra, India

Kharanshu Bojak
Department of Metallurgical Engineering and Material Science
IIT Bombay
Mumbai, Maharashtra, India

Ananya Burli
Department of Chemical Engineering
IIT Bombay
Mumbai, Maharashtra, India

Manisha Choudhury
Department of Biosciences and Bioengineering
IIT Bombay
Mumbai, Maharashtra, India

Susmita Ghosh
Department of Biosciences and Bioengineering
IIT Bombay
Mumbai, Maharashtra, India

Abhiram Gokhale
Department of Biosciences and Bioengineering
IIT Bombay
Mumbai, Maharashtra, India

Ankit Halder
Department of Biosciences and Bioengineering
IIT Bombay
Mumbai, Maharashtra, India

Ashwini Khaladkar
Department of Biosciences and Bioengineering
IIT Bombay
Mumbai, Maharashtra, India

Harsh Khatri
Department of Biosciences and Bioengineering
IIT Bombay
Mumbai, Maharashtra, India

Gaurish Loya
Department of Chemical Engineering
IIT Bombay
Mumbai, Maharashtra, India

Krishi Mantri
Department of Chemical Engineering
IIT Bombay
Mumbai, Maharashtra, India

Amrita Mukherjee
Department of Biosciences and Bioengineering
IIT Bombay
Mumbai, Maharashtra, India

Mehar Un Nissa
Department of Biosciences and Bioengineering
IIT Bombay
Mumbai, Maharashtra, India

Medha Gayathri J. Pai
Department of Biosciences and Bioengineering
IIT Bombay
Mumbai, Maharashtra, India

Viswanthram Palanivel
Department of Biosciences and Bioengineering
IIT Bombay
Mumbai, Maharashtra, India

Shashwati Parihari
Department of Biosciences and Bioengineering
IIT Bombay
Mumbai, Maharashtra, India

Firuza Parikh
Department of Assisted Reproduction and Genetics
Jaslok Hospital and Research Centre
Mumbai, Maharashtra, India

Jing Ren
State Key Laboratory of Proteomics
Beijing Proteome Research Center
National Center for Protein Sciences
Beijing, China

Jyotirmoy Roy
Department of Chemical Engineering
IIT Bombay
Mumbai, Maharashtra, India

Akanksha Salkar
Department of Biosciences and Bioengineering
IIT Bombay
Mumbai, Maharashtra, India

Rhythm Shah
Department of Metallurgical Engineering and Material Science
IIT Bombay
Mumbai, Maharashtra, India

Om Shrivastav
Kasturba Hospital for Infectious Diseases
Mumbai, Maharashtra, India

Alisha Srivastava
Department of Genetics
University of Delhi
New Delhi, Delhi, India

Sanjeeva Srivastava
Department of Biosciences and Bioengineering
IIT Bombay
Mumbai, Maharashtra, India

Kruthi Suvarna
Department of Biosciences and Bioengineering
IIT Bombay
Mumbai, Maharashtra, India

Ayushi Verma
Department of Biosciences and Bioengineering
IIT Bombay
Mumbai, Maharashtra, India

Radha Yadav
Department of Mathematics
IIT Bombay
Mumbai, Maharashtra, India

Xiaobo Yu
State Key Laboratory of Proteomics
Beijing Proteome Research Center
National Center for Protein Sciences
Beijing, China

1

Clinical and epidemiological context of COVID-19

Viswanthram Palanivel, Akanksha Salkar, Radha Yadav, Renuka Bankar, Om Shrivastav, and Arup Acharjee

CONTENTS

DOI: 10.1201/9781003220787-1

List of abbreviations

BiPAP – Bilevel positive airway pressure
CFR – Case fatality rate
COVID-19 – Coronavirus Disease 2019
CPAP – Continuous positive airway pressure
GISAID – Global Initiative on Sharing All Influenza Data
NIV – Non-invasive ventilation
WHO – World Health Organization

1.1 Introduction

1.1.1 COVID-19: A 21st-century mayhem

SARS-CoV-2 has caused a pandemic of unimaginable proportions. It was first identified in China in December 2019 and was named COVID-19 in February 2020 by WHO (Carvalho, Krammer, and Iwasaki 2021). Less than a month after it was named, COVID-19 was officially declared a global pandemic by WHO in March 2020 (Carvalho, Krammer, and Iwasaki 2021). This viral disease is transmitted via droplets and initially presents itself as a lower respiratory tract infection. Most infected individuals manifest mild disease and present flu-like symptoms such as dry cough, body ache, and fever. Atypical symptoms such as anorexia, diarrhea, and vomiting have also been reported. Some patients manifest severe respiratory illness with conditions such as dyspnea and hypoxia or they would show greater than 50% lung involvement on imaging and may further progress to the critical stage with acute respiratory distress syndrome (ARDS), respiratory failure, viremia, and eventually multi-organ system dysfunction (Zaim et al. 2020). Extrapulmonary involvement of the virus due to angiotensin-converting enzyme (ACE2) receptor expression in different organs increases the complexity of disease pathogenesis (Dong et al. 2020). The infected individuals were initially treated with a combination of antiviral or antibiotics with other supplementary drugs to alleviate symptoms due to the lack of specific COVID-19 medicines and vaccines. The pandemic severely impacted the health systems and economies globally, thus it's been called a "21st-century mayhem".

1.1.2 An overview of SARS-CoV-2

SARS-CoV-2 is known to be the seventh coronavirus to infect humans and third among the beta coronaviruses to cause a major outbreak in this century (Su et al. 2016; Carvalho et al. 2021). It is named after the SARS-CoV as their genomes are 80% similar to each other (Zhu et al. 2020), and the genetic arrangement of open reading frames (ORFs) also resembles the Middle East respiratory syndrome coronavirus (MERS-CoV) strain (Naqvi et al. 2020). The virus is comparatively larger, with 60–140 nm in diameter, and has a 9–12 nm spike surrounding it. SARS-CoV-2 is an enveloped single-stranded RNA virus known to be one of the giant viral genomes with 26–32 kb in length (Naqvi et al. 2020). The genetic constitution of SARS-CoV-2 contains 13–15 ORFs with around 30,000 nucleotides. The viral genome holds 11 coding genes expressing 12 viral proteins (Kronbichler et al. 2020). Structurally, the virus has four structural proteins: spike (S), membrane (M), envelope (E), and nucleocapsid (N) proteins. The spike protein (S) binds with host cell receptor ACE-2 found in many human organs like lungs, heart, intestine, and kidneys, allowing the virion to penetrate the cell. Inside the cell, the virus replicates and invades the cell machinery, disturbing its homeostasis, leading to apoptosis. Thus, these apoptotic cells and their fragments accumulate as debris and fluid in the lungs, which congest the airway, causing breathing distress and pneumonia in the patients. During the initial phase of the infection, the innate immune response acts as the first-line defense by producing important cytokines and further initiates adaptive response to take over (Naqvi et al. 2020).

Moreover, the host immune system exhibits a hyperactive response by attacking the infected and neighboring healthy cells, leading to complete respiratory failure and death. In rare cases, virions enter the bloodstream causing multiple organ failures. The fatality rate of COVID-19-infected patients ranges from 2% to 3%, and for the hospitalized patient, it ranges from 4% to 11% (Baric 2020). Molecular testing techniques such as reverse transcriptase-polymerase chain reaction (RT-PCR) are considered a confirmatory test to detect SARS-CoV-2 in the throat and nasopharyngeal swabs. The viral RNA is also detected in stool samples of infected patients, sometimes at higher levels, inferring the possibility of fecal–oral transmission (Mesoraca et al. 2020). During the current pandemic, there has been unparalleled contribution made by the scientific community in understanding the molecular mechanism behind the clinical manifestations for developing tools for early diagnosis and prognosis (Aggarwal et al. 2021; Bankar et al. 2021; Suvarna et al. 2021; Messner et al. 2020; Shen et al. 2020; Su et al. 2020). Furthermore, these efforts will help us develop new drugs targeting SARS-CoV-2.

1.2 Clinical manifestation of COVID-19

Understanding of COVID-19 characteristics is evolving over the period as the mutant strains are rising in several areas. COVID-19-infected patients have a broad spectrum of clinical manifestations and seem to predict a favorable trend in most patients. Pneumonia is a pervasive manifestation of SARS, MERS, and COVID-19. Most of the clinical features of SARS-CoV-2 are similar to the characteristics of the severe acute respiratory syndrome (SARS). They include fever, cough, general weakness, body ache, diarrhea, vomiting, myalgia, and dyspnea with progression to pneumonia and acute respiratory distress syndrome (Huang et al. 2020; Wang et al. 2020). Figure 1.1 represents the clinical manifestations of COVID-19 and its timeline of symptom relations from asymptomatic to symptomatic state. The incubation period of SARS-CoV-2 from the exposure of the virus to exhibiting clinical signs is between 5 and 14 days (Guan et al. 2020). Besides respiratory distress and failure, infected patients are also known to manifest viremia, which causes multi-organ failure putting the patient at higher risk and eventually leading to death. Global expansion of COVID-19 disease with a substantial rise in the number of positive patients also annotates other new symptoms arising. These clinical manifestations differ from patient to patient based on certain risk factors like age, health conditions, and gender, among others (Yadav 2021). Among these, specific age groups

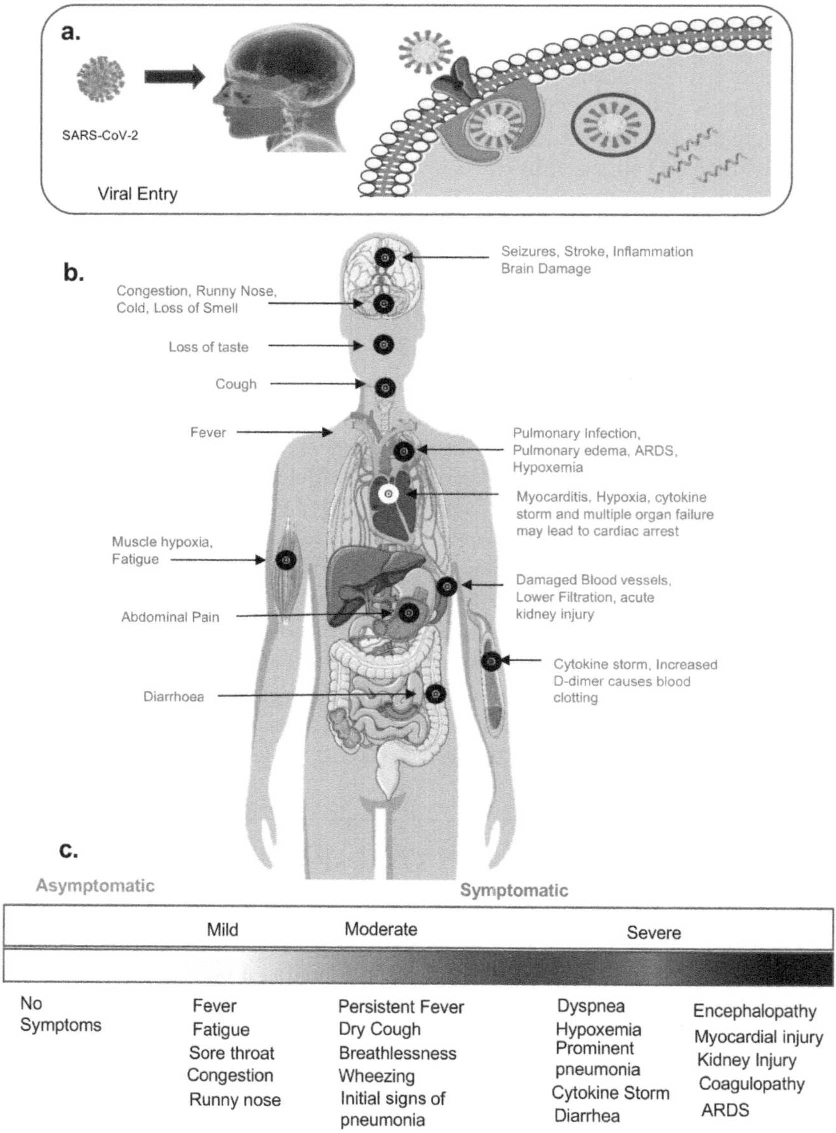

FIGURE 1.1 Schematic representation showing the impact of SARS-CoV-2 in human body. (a) Entry of SARS-CoV-2 into the human body through nasal cavity. (b) Pathophysiology of COVID-19: complications seen in patients with COVID-19 patients. (c) A timeline of symptoms in COVID-19 patients.

Source: This figure was created using Servier Medical Art templates, which are licensed under a Creative Commons Attribution 3.0 Unported License; https://smart.servier.com.

have also been noted as a risk factor for the severity of the disease. At the same time, the younger population is studied to be with mild symptoms or asymptomatic, which might be crucial in further spreading the disease (Kronbichler et al. 2020). Severe cases are often treated in an intensive care unit (ICU) and require artificial ventilation as a breathing aid. The case fatality rate (CFR) of COVID-19 is estimated between 3.4% and 11% (Novosad et al. 2020).

1.2.1 Asymptomatic patients

Asymptomatic infection is when a patient develops no symptoms such as fever, cough, or any other condition during illness, and they seem to constitute approximately 40–45% of SARS-CoV-2 infections (Oran and Topol 2020; Gao et al. 2021). Asymptomatic infections have no notable incubation period due to the absence of clinical symptoms; this indicates that the transmission is very unpredictable and could even occur earlier in the course of infection. However, asymptomatic patients also have few objective clinical abnormalities, which are very subtle to be identified. In a study of 24 asymptomatic patients, chest computed tomography (CT) revealed typical ground-glass opacities or patchy shadowing in over 50% of patients. Besides, another 20% had atypical imaging abnormalities (Hu et al. 2020). Around five patients developed subtle symptoms typical in other COVID-19 patients with low-grade fever a few days after diagnosis. In another study, 55 asymptomatic patients were identified by tracing close contacts; 67% had CT evidence of pneumonia since admission. Besides, only two had developed hypoxia, and the rest of them were recovered (Wang et al. 2020). These findings suggest CT imaging's role in diagnosing COVID-19, even for asymptomatic patients in the absence of RT-PCR-based testing.

Asymptomatic patients are considered viable carriers; and these patients can transmit the SARS-CoV-2 for a while until they start exhibiting symptoms and get diagnosed (Gao et al. 2021). Some patients with standard radiology and no symptoms also appeared to be symptomatic with the follow-up period (i.e., they were presymptomatic). Their laboratory findings revealed that creatinine and D-dimer parameters were elevated (Kronbichler et al. 2020). However, studies have shown that the viral load detected in patients with the asymptomatic condition is similar to other symptomatic patients (Gao et al. 2021). Most asymptomatic COVID-19 cases are seen in younger age groups and women; others seem to develop symptoms during the incubation period (Kronbichler et al. 2020). Some younger cohorts with asymptomatic SARS-CoV-2 infection also seem to develop symptoms at follow-up, but they have an excellent recovery prognosis (Gao et al. 2021). These findings suggest that asymptomatic patients should not be neglected during contact tracing or during mass testing to stop the reemergence of the virus.

1.2.2 Symptomatic patients

The spectrum of clinical manifestations of COVID-19 patients with symptomatic condition ranges from mild to critical condition; many patients are found to be grouped under non-severe categories. The most frequently mentioned factors increasing the risk for severe infection are age (>60 years), male sex, and postmenopausal female (Wolff et al. 2020). Symptomatic patients are classified into mild, moderate, severe, and critical. The clinical symptoms exhibited by the COVID-19 patients of different conditions are as follows (Wolff et al. 2020):

- *Mild:* Fever, cough, sore throat, fatigue, muscle pain, anorexia, anosmia, body ache, and headache
- *Moderate:* Clinical features of mild condition with breathlessness, mild pneumonia manifestation, and chest imaging findings
- *Severe:* Respiratory infection with shortness of breath, respiratory rate ≥30 breaths per minute, oxygen saturation (SpO_2) ≤93% in room air, and severe lesions shown in chest imaging
- *Critical:* Respiratory failure, a substantial decrease in oxygen saturation level (need mechanical ventilation to breath), septic shock, and multiple organ failure will be seen in the latent stage, which requires ICU monitoring treatment.

Most of the world population infected with SARS-CoV-2 has mild symptoms, and many seem to recover in an earlier stage of infection. Patients with mild infection experience primary symptoms like fever, cough, general weakness, and body ache. Some mild patients also experience breathlessness with oxygen saturation (SpO_2) of 92–94%, which usually relapses to normal (Wolff et al. 2020). Certain underlying medical conditions, which act as a risk factor, promote this non-severe to severe transition. Some of the most common clinical parameters associated with severity are increased creatinine level, increased interleukin-6 (IL-6) and C-reactive protein (CRP), increased procalcitonin level (PCT), increased blood urea nitrogen (BUN), and decreased lymphocyte count. Other clinical parameters involved with the fatal condition are increased neutrophils, leukocytes, and D-dimer levels. Patients with severe infection have an oxygen saturation of less than 88%, making them utterly dependent on oxygen supplementation as their life support (Wolff et al. 2020). In many cases, the host immune response against the virus also kills the infected alveolar cells causing acute respiratory distress syndrome in critical cases. Elevated levels of cytokines such as IL-6 and CRP act as clinical markers showing the elevated immune response against virus-infected cells.

1.3 Management of COVID-19 pandemic

1.3.1 Prophylaxis

The Centers for Disease Control and Prevention (CDC) (Guidance for COVID-19, CDC) has suggested a list of guidelines to limit the spread of person-to-person transmission of the virus. The guidelines of CDC are as follows: (i) To maintain 6 feet physical distance between individuals, (ii) to avoid socializing in public places such as restaurants, bars, fitness centers, theaters, and social gatherings, (iii) to avoid being in poorly ventilated spaces, (iv) to wear a mask or any other protection covering the nose and mouth properly, (v) often washing or sanitizing hands by using alcohol-based sanitizer or hand rub or soap water, (vi) to clean and disinfect the surfaces regularly with sodium hypochlorite or any other disinfectant, (vii) covering nose and mouth with a tissue while sneezing and coughing, (viii) to immediately seek medical attention if having flu-like symptoms, and (ix) to self-isolate or quarantine person with clinical symptoms of COVID-19. Following these guidelines can be an excellent preventive measure for people during these pandemic times. SARS-CoV-2 is highly sensitive to chemicals such as ethanol (75%), ether, and chlorine disinfectants (sodium hypochlorite) and also can be inactivated by ultraviolet radiation and heat treatment at 56°C for 30 minutes (Hoffmann et al. 2021; Li et al. 2020). These measures have been proven to be effective in curbing the spread of SARS-CoV-2.

1.3.2 Treatment

Oxygen therapy

Breathlessness and hypoxemia are considered as primary clinical features in COVID-19 infection. Lungs have a high expression of ACE2. In a disease progression to severity, pro-inflammatory cytokines such as interleukin-6 and interleukin-8 and other signaling molecules like monocyte chemoattractant protein-1 (MCP-1), tumor necrosis factor-alpha (TNF-α), and granulocyte colony-stimulating factor (G-CSF) are released, attracting neutrophils and T cells that induce lung injury leading to ARDS. Hence, mechanical oxygen support is essential in COVID-19 management. Non-invasive ventilation (NIV) like continuous positive airway pressure (CPAP), bilevel positive airway pressure (BiPAP), and hyperbaric oxygen therapy (HBOT) are widely used. Similarly, invasive ventilation such as endotracheal/orotracheal tube and extracorporeal membrane oxygenation (ECMO) is preferred by doctors in critical conditions.

Antiviral therapy

Some of the antiviral therapies which physicians predominantly use are as follows:

i) Targeting viral entry (fusion inhibitors)
   - Baricitinib, a potent Janus kinase (JAK) inhibitor, binds to AP2-associated protein kinase 1. It inhibits both host inflammatory response and viral entry (Zhang et al. 2020).
   - Camostat mesylate is a TMPRSS2 protease inhibitor that blocks the SARS-CoV-2 infection of lung cells (Hoffmann et al. 2021).

ii) Targeting virus replication
   - *Reverse transcription inhibitors:* Remdesivir is an RNA-dependent RNA polymerase (RdRp) inhibitor and is a broad-spectrum antiviral drug against the significant class of single-stranded RNA (ssRNA) viruses (Amirian and Levy 2020).
   - *Nucleotide analogs:* Favipiravir is a purine (Guanine) analog. It blocks the RdRp to arrest viral RNA synthesis (Furuta, Komeno, and Nakamura 2017). Ribavirin is a guanine derivative analog. It destabilizes viral RNA by inhibiting RNA capping.

iii) Protease inhibitors

- *Lopinavir:* Lopinavir inhibits the main protease (MPro) of SARS-CoV-2. It is used as a lopinavir–ritonavir combination, which shows high efficiency. Primarily, it is an HIV antiviral drug (Nutho et al. 2020).

Passive immunization

i) Plasma therapy

Passive virus neutralization by antibodies generated in other patients is the sole principle behind plasma therapy. Convalescent plasma is collected from COVID-19 individuals who have recovered at least three weeks before and used to immunize severely or critically ill patients passively. There was no literature available for antiviral treatment or prevention of infection for the novel coronavirus; therefore, passive immunization was the first therapeutic choice. A study (Cunningham, Goh, and Koh 2020) states that immunoglobulin therapy to severe patients helps in decreasing viral load inflammation and boost oxygenation in ten critically ill patients.

ii) Antibody therapy

The receptor-binding domain (RBD) of spike protein (S) shows a strong affinity to the receptor on human cells, angiotensin-converting enzyme 2 (ACE2). Type 2 alveolar cells of lungs, stratified epithelium, myocardial cells, and the proximal tubule cells in the kidney are marked as high risk for sites for COVID-19 infection due to high ACE2 expression. Anti-S1 human monoclonal antibody 80R is potent in neutralizing SARS-CoV-2 by blocking the binding of S1 to ACE2 (Tai et al. 2020).

iii) Immunomodulatory drugs/anti-inflammatory drugs

Some of the commonly used immunomodulatory drugs are as follows:

- *IL-6 inhibitors:* Tocilizumab is a humanized monoclonal antibody against IL-6 and is widely used to treat severe and critical patients.
- *IL-1 inhibitor:* Anakinra is used as an IL-1 inhibitor.
- *TNF-alpha cytokine inhibitors:* Adalimumab, etanercept, and infliximab.
- Corticosteroids reduce the activation of inflammatory mediators, leading to the expression of anti-inflammatory genes that inhibit nuclear factor-kappa B (NF-κB) and interleukins IL-4, IL-10, and IL-13.

1.4 Comorbidities and their risk factors in severity progression

SARS-CoV-2 infection appears to vary with the underlying medical conditions of infected patients. The presence of medical conditions like diabetes mellitus, hypertension, cardiovascular diseases, chronic liver disease, chronic kidney disease, chronic neurological disease, cancers with and without direct immunosuppression, acquired immunodeficiency syndrome, tuberculosis, sickle cell anemic disorders, and alcoholism reportedly affect the COVID-19 prognosis (Clark et al. 2020). The following points explain the correlation of a few common risk factors involved with the severity in most of the COVID-19 patients.

Diabetes mellitus

Many claims have been made regarding mechanisms aggravating the effect of diabetes on COVID-19. They include poor hyperglycemic control, impaired immune response after the viral entry into the cell, and diabetes-related complications. Studies suggest that the severity of the disease is directly related to hyperglycemic conditions, as there is a lower rate of deaths recorded in individuals with controlled glucose (Lim et al. 2021). However, in this outline, the possible diabetogenic effect of SARS-CoV-2 should be considered as it directly acts on crucial metabolic organs, including the β-cells. It results in the development of hyperglycemia or sudden deterioration of preexisting diabetes (Rubino et al. 2020). Diabetes also favors the SARS-CoV-2 infection by compromising the adaptive immune host response against the virus rather than increasing the innate immune inflammatory reaction. The negative effect of COVID-19 on people with diabetes can also be facilitated by diabetes-related complications such as obesity, hyperlipidemia, and chronic kidney disease. These factors can be associated with the poor prognosis of the COVID-19 patients despite treatment (Lim et al. 2021).

Hypertension

It is ambiguous whether abnormal blood pressure is a risk factor for severity transition in COVID-19 patients or the antihypertensive medications responsible for maintaining blood pressure are or not less of a risk factor. However, studies suggest that controlled blood pressure can be a vital consideration to reduce the burden of the disease, and it also alters the susceptibility of the SARS-CoV-2 viral infection (Yadav 2021). Hypertensive patients are often treated with angiotensin receptor blockers (ARBs) and ACE inhibitors, commonly referred to as renin-angiotensin-aldosterone inhibitors. These promote ACE2 expression, which is the entry point of the SARS-CoV-2 to enter into the lung cells.

1.5 Global scenario

The dataset for COVID-19 and key sustainable development indicators is sourced from *Our World in Data* for the purpose of this analysis on May 2, 2021 (Ritchie et al. 2020). Out of the 7.8 billion people globally, a total of 152,870,507 (2%) people tested positive for COVID-19. As of May 2021, 29.5% of all positive cases globally came from Europe, followed by Asia (26.5%) and North America (24.6%). Europe also accounted for the greatest number of deaths globally (31.9%), followed by North America (26.5%) and South America (21.2%). Although Asia constitutes 59.5% of the world population, it accounted for a significantly lower number of deaths worldwide (16.6%) with a total of 8,728 positive cases per million and 114 deaths per million. South America had the highest number of deaths per million

(58,219 cases per million, 1,575 deaths per million). In contrast, North America had the highest number of positive cases per million (63,551 cases per million, 1,434 deaths per million), followed by Europe (60,169 cases per million, 1,363 deaths per million). Of all continents, Oceania and Africa, which together constitute 17.7% of the global population had the least number of positive cases (0.03% and 3% of all global cases, respectively), number of positive cases per million (1,026 and 3,410 cases per million, respectively), number of deaths (0.03% and 3.8% of all deaths worldwide, respectively) and number of deaths per million (24 and 91 deaths per million, respectively). Higher income countries in North America and Europe are leading the vaccination efforts with 29.8%, and 21.8% of their population vaccinated with at least one dose of vaccine, respectively. The rates are low for Asia, standing at just 4.3%.

India saw an increasing number of new cases till September 2020, which gradually declined for the next six months but later witnessed a massive spike in the COVID-19 positive cases in April 2021, which drove its share of total global cases up by 3.6% points from 9.5% on March 31, 2021, and that of total global deaths up by 1.1% points from 5.8% on March 31, 2021 (Figures 1.2 and 1.3) (WHO Coronavirus (COVID-19) Dashboard 2021; Ritchie et al. 2020).

Out of the world's top 100 most populous countries, the 10 countries that had the highest number of positive cases per million and had very high (>1,000) deaths per million were Czechia, the United States, Sweden, the Netherlands, Belgium, France, Portugal, Hungary, Switzerland, and Spain. Interestingly, these countries had the following common global indicators (mean ± standard deviation): high median age (42.78 ± 2.23), a high percentage of the population aged 65 or older

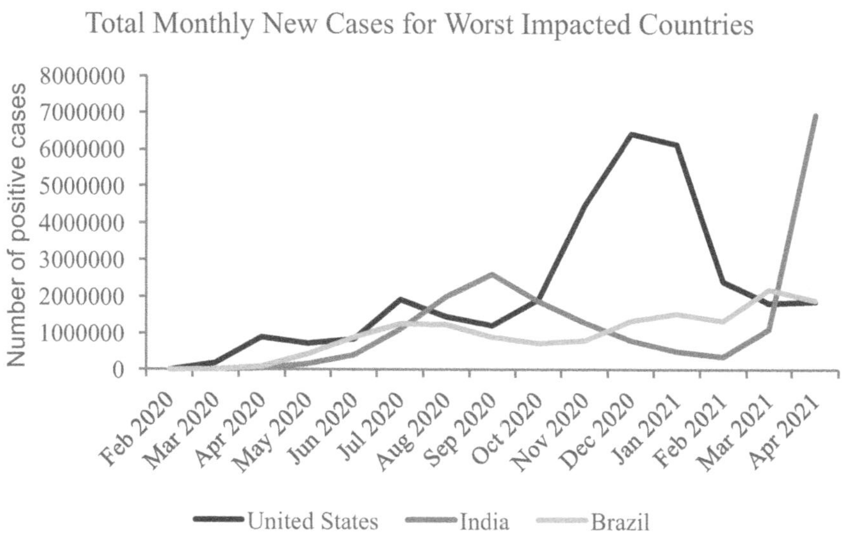

FIGURE 1.2 Total monthly new cases for worst impacted countries: The United States, India, and Brazil.

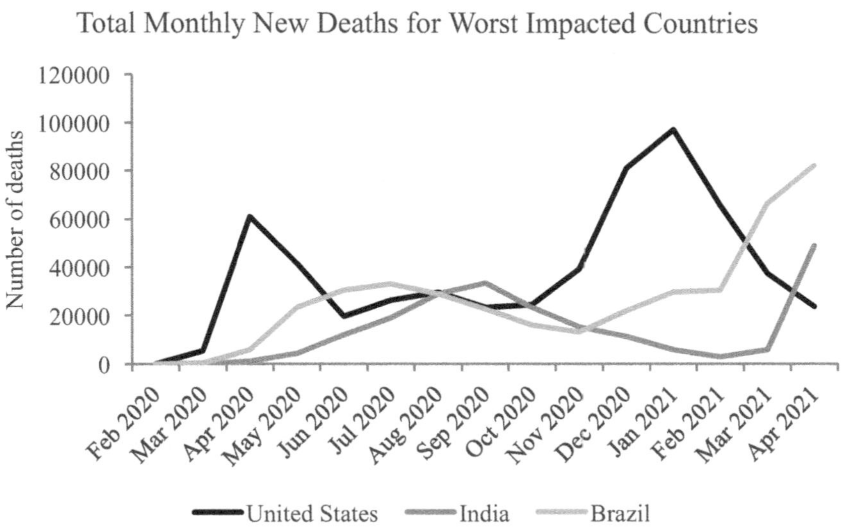

FIGURE 1.3 Total monthly new deaths for worst impacted countries: The United States, India, and Brazil.

(18.94% ± 1.55%), a high percentage of the population aged 75 or older (12.59% ± 1.41%), low percentage of the population with extreme poverty (0.39% ± 0.44%), low to moderate cardiovascular death rate (142.82 ± 62.14), low diabetes prevalence (6.69% ± 2.21%), a high percentage of female smokers (24.11% ± 4.86%), high percentage of male smokers (30.12% ± 5.66%), high life expectancy (81.39 ± 2.26 years), and very high HDI (0.91 ± 0.03).

1.6 COVID-19 pandemic in India

The dataset for COVID-19 and key sustainable development indicators for India is sourced from *Our World in Data* for the purpose of this analysis on May 2, 2021 (Ritchie et al. 2020). As of May 2, 2021, India has 19,919,004 positive cases, 16,281,710 recovered cases, and 218,951 deaths. Among all Indian states, Maharashtra was impacted the most with 31.7% of nationwide deaths and 23.3% of total nationwide positive cases, followed by Delhi (7.8% deaths and 6% positive cases), Karnataka (7.3% deaths and 8.2% positive cases), Tamil Nadu (6.5% deaths and 6% positive cases), and Uttar Pradesh (6.1% deaths and 6.6% positive cases). A unique common indicator for these five states is that more than 40% of their population have 0–1 sleep rooms, and 35–40% have two rooms, which are significantly higher percentages than other less impacted states. Kerala, on the other hand, had the second-highest number of positive cases in India (8.2% of total cases in India) but significantly lower death cases (2.4% of total deaths in India) (Figures 1.4 and 1.5) (WHO Coronavirus (COVID-19) Dashboard 2021; Ritchie et al. 2020). One potential reason for the increased number of positive cases in April 2021

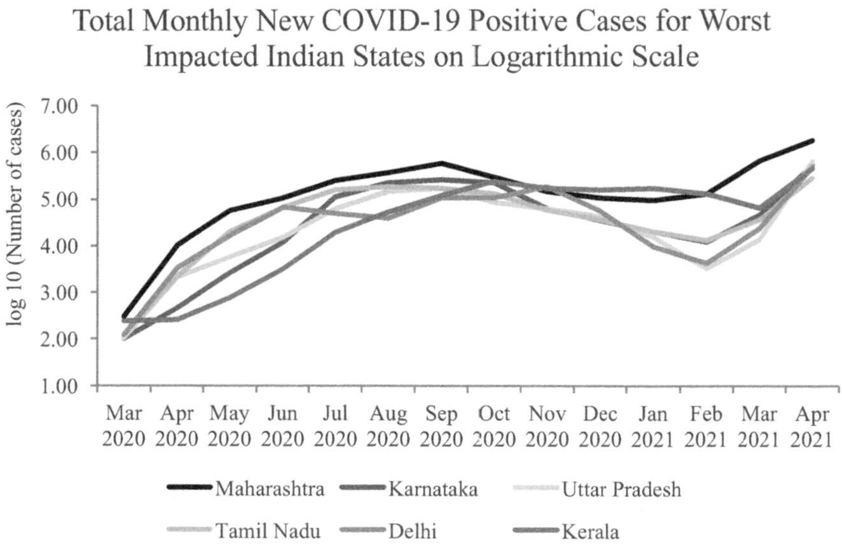

FIGURE 1.4 Total monthly new COVID-19 positive cases for worst impacted Indian states.

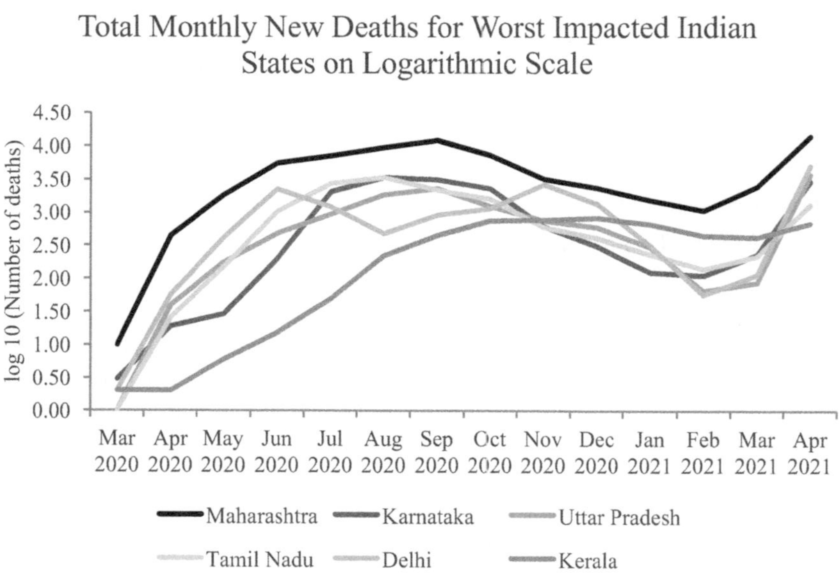

FIGURE 1.5 Total monthly new deaths due to COVID-19 for worst impacted Indian states.

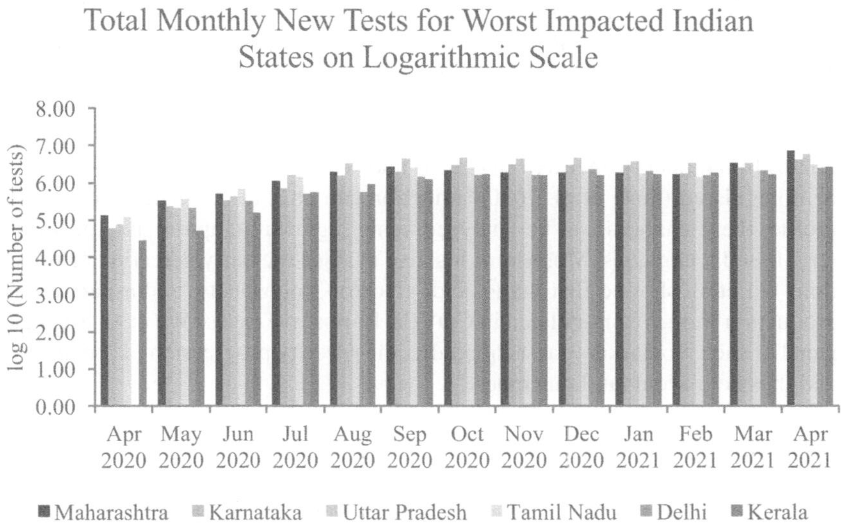

FIGURE 1.6 Total monthly new tests for worst impacted Indian states.

could be the decreased or flattened number of COVID-19 tests carried out between December 2020 and February 2021 (Figure 1.6). (WHO Coronavirus (COVID-19) Dashboard 2021; Ritchie et al. 2020). The emergence of newer strains of SARS-CoV-2 might have contributed to the spike as well.

1.7 Authors' perspective: Omics technologies for COVID-19 research

1.7.1 Early-phase markers for diagnosis

RT-PCR test detects the presence of viral genetic material, and it is a highly sensitive test that is considered the benchmark for COVID-19 detection; however, the false-negative rate of this test is strikingly high (Kucirka et al. 2020). Antigen test can be carried out very rapidly but has very low sensitivity (Scohy et al. 2020). Both RT-PCR and antigen tests validate COVID-19 diagnosis only in later stages of infection, leading to the spread of COVID-19 in the community. Mass spectrometry–based techniques such as proteomics and metabolomics are highly sensitive and give a descriptive idea about the patient's severity state and other medical conditions (Aggarwal et al. 2021). Several biological pathways also get hampered during the infection process. Intermediate biomolecules which are perturbed can be analyzed for alterations in their levels. Omics technologies such as proteomics, metabolomics, and lipidomics can be used to provide information about alteration in biomolecules during an early infection phase. All such biomolecules are categorized as biomarkers or early-phase markers. These biomarkers can help in the proper isolation of the patient and the prognosis of severity. Proteomics analysis of nasopharyngeal swabs showed an upregulated level of lactate dehydrogenase A, lactate dehydrogenase B, and interleukin-6 in COVID-19 patients (Bankar et al. 2021). Downregulation of APOA1, APOA2, and APOM was observed in COVID-19 patients (Shen et al. 2020). Inflammatory markers like IL-6 and C-reactive protein were shown to be upregulated (Song et al. 2020). Infection can change the metabolic or lipid profile; these changes can be analyzed using high-throughput mass spectrometry–based techniques. Moreover, this also enables the early detection of infection. Wu et al. showed that malic acid, glycerol-3-phosphate, and aspartic acid production were downregulated in COVID-19-positive cases (Wu et al. 2020). A panel of such biomarkers can help predict the severity progression of the disease and thus help avoid fatal outcomes.

1.7.2 Prognosis of severity

One of the primary causes of fatality for COVID-19 patients is the lack of timely prognosis. Timely intervention, especially for patients moving from mild to severe form of COVID-19, can help assuage the symptoms leading to better recovery. To date, most of the factors that influence stratification of COVID-19 patients include hematological laboratory results for plasma cytokine levels, CRP, leukocyte, lymphocyte, D-dimer, erythrocyte sedimentation rate, and other indicators like the ratio of CD4+ and CD8+ T cells, change in the number of platelets, platelet-to-lymphocyte ratio, and acute kidney injury.

Moreover, clinicians take help from radiological findings to assess the spread of infection in the internal organs, such as pleural and pericardial effusion, diffuse alveolar damage pattern, bronchial wall thickening, enlargement of lymph node, and opacities in the lungs. However, as evident, these changes happen only after a severe bout of infection sets in, making clinical interventions difficult and perhaps less efficacious.

Therefore, the lack of accurate biomarkers of severity is a concern. Omics studies have shown that, indeed, the severity of the patient can be predicted using plasma biomarkers (Aggarwal et al. 2021). In a landmark study, Shen et al. employed a random forest machine learning algorithm on the proteomic and metabolomic data obtained from non-severe and severe cohorts. They listed at least 22 proteins and seven metabolites that could predict the severity of the disease (Shen et al. 2020). Similarly, another study identified that the exosomes during severe disease were progressively enriched in a metabolite monosialodihexosyl ganglioside (Song et al. 2020). In contrast, another plasma multi-omics study (Su et al. 2020) of 139 COVID-19 patients representing both non-severe and severe conditions was found to be dependent on higher circulating cytokines and identified the downregulated metabolites and metabolic processes. Further, the multi-omic profiles from their plasma samples showed a sharp transformation of the metabolic profile from mild to moderate progression, as shown by the loss of amino acids, lipids, and xenobiotic metabolism, with a substantial increase in inflammatory cytokines.

1.7.3 Therapeutics

A large scale of multi-omics studies on COVID-19 is being carried out to explore the therapeutic options in treating SARS-CoV-2 infection. Transcriptomics analysis provides broad knowledge about the expression and activity of the viral genes during infection. These could help identify the potential therapeutic drug to target these expressed proteins, which causes severity in the patients. Proteomics investigation can also provide information about the epigenetic and post-translational modifications in host cells caused by the SARS-CoV-2 infection. Viruses regulate the host cellular factors for its replication inside the cell, and these actions could act as a factor in epigenetic regulations and posttranslational modifications. Thus, the proteomic approaches can differentiate the normal host proteins from virus-infected cellular proteins, and these could serve as potential biomarkers for therapeutic and diagnostic purposes.

1.7.4 Vaccine

As the COVID-19 situation worsens day by day, the need to design an optimum vaccine with higher efficacy is growing. Although multiple vaccines are available on the market, their efficacy varies among countries making them less potent. Multi-omics technologies can be used to find suitable epitopes which can produce enough immune response against SARS-CoV-2. Using a proteomic investigation of the viral components, suitable viral peptides and viral protein sequences can be explored thoroughly using multiple data analysis software for identifying effective vaccines. The genomic approach by sequencing RNA can help scientists work on the genetic expression level to identify the potential candidate for vaccine development.

1.7.5 Genomic surveillance

Genomic epidemiology has come of age during the current COVID-19. The pandemic has also ushered scientists in tracking genomic changes to a virus at a speed and scale never seen before (Cyranoski 2021). The coronavirus has a mutation rate that is tenfold lower than that of other RNA viruses, owing to the proofreading activity of nsp14. It is known to have a mutation rate of about 8×10^{-4} nucleotides/genome per year. The researchers have used these mutations to stratify the virus to a particular lineage or clade. Global Initiative on Sharing All Influenza Data (GISAID) hosts the largest curated international repositories of SARS-CoV-2 sequence data (Hemrajata 2021). Around 3,913 genomes have been collected and uploaded in GSAID SARS-CoV-2 Genomic Epidemiology (EpiCoV) platform between September 2020 and April 2021. This data sheds light on strain diversity and putative intra- and intercontinental transmissions by providing a concurrent summary of the distribution of SARS-CoV-2 clades across geographical reasons. The United Kingdom and Denmark constitute 45% and 7% of genomes on the database (COG-UK 2020; Cyranoski 2021). One such study undertaken in North California revealed that WA1 strain associated with Washington state and six other strains were introduced cryptically into North

California (Deng et al. 2020). Another study from the Netherlands demonstrated how a combination of WGS and epidemiology helped understand the transmission and strengthen the evidence base for public health decision-making to implement strict measures (Oude Munnink et al. 2020). In another study from Japan, genome sequencing was performed for returnees or travelers arriving in Japan at airport quarantining stations (Sekizuka et al. 2021). The results demonstrated that testing and genome sequencing should be performed efficiently to monitor the introduction of new strains into the community. Sequencing for Public Health Emergency Response, Epidemiology and Surveillance (SPHERES), a SARS-CoV-2 sequencing initiative, is established under CDC's advanced molecular detection program. It aims at providing sequence data in real time for investigating COVID-19 cases and clusters. Despite these global efforts, there are still some gaps in the surveillance that need to be filled. These can be due to the lack of national-level consortium in many countries that have intense transmission.

1.8 Conclusions

Proper understanding of the clinical manifestation of the disease, biomarkers of infection, and its severity, and the epidemiology and transmission dynamics plays an essential role in curbing any disease. Considering a gargantuan pandemic like COVID-19, understanding these becomes crucial. The scientific community has made an unparalleled contribution in increasing awareness about the clinical features of COVID-19. Many countries have established surveillance networks for studying the transmission dynamics of the virus. There has been an increase in the use of "omics" technologies for understanding the SARS-CoV-2 and for surveillance of the virus. Vaccines have come in a record time. Although much has been achieved, the war against COVID-19 is far from over. The new waves of infection, cases of reinfection, and mutant strains have revealed the chinks in our armor, and these gaps need to be addressed in the times to come. Potential way forward will be to undertake more collaborative research that is not restricted to a country but is rather extended worldwide.

REFERENCES

Aggarwal, Shalini, Arup Acharjee, Amrita Mukherjee, Mark S. Baker, and Sanjeeva Srivastava. 2021. "Role of Multiomics Data to Understand Host—Pathogen Interactions in COVID-19 Pathogenesis." *Journal of Proteome Research* 20 (2). American Chemical Society: 1107–32. https://doi.org/10.1021/acs.jproteome.0c00771.

Amirian, E. Susan, and Julie K. Levy. 2020. "Current Knowledge about the Antivirals Remdesivir (GS-5734) and GS-441524 as Therapeutic Options for Coronaviruses." *One Health* 9 (June): 100128. https://doi.org/10.1016/j.onehlt.2020.100128.

Bankar, Renuka, Kruthi Suvarna, Saicharan Ghantasala, Arghya Banerjee, Deeptarup Biswas, Manisha Choudhury, P. Viswanthram, et al. 2021. "Proteomic Investigation Reveals Dominant Alterations of Neutrophil Degranulation and MRNA Translation Pathways in COVID-19 Patients." *Iscience* 24 (3). Elsevier. https://doi.org/10.1016/j.isci.2021.102135.

Baric, Ralph S. 2020. "Emergence of a Highly Fit SARS-CoV-2 Variant." *New England Journal of Medicine* 383 (27). Massachusetts Medical Society: 2684–86. https://doi.org/10.1056/NEJMcibr2032888.

Carvalho, Thiago, Florian Krammer, and Akiko Iwasaki. 2021. "The First 12 Months of COVID-19: A Timeline of Immunological Insights." *Nature Reviews Immunology*, March. Nature Publishing Group: 1–12. https://doi.org/10.1038/s41577-021-00522-1.

Clark, Andrew, Mark Jit, Charlotte Warren-Gash, Bruce Guthrie, Harry H. X. Wang, Stewart W. Mercer, Colin Sanderson, et al. 2020. "Global, Regional, and National Estimates of the Population at Increased Risk of Severe COVID-19 Due to Underlying Health Conditions in 2020: A Modelling Study." *The Lancet Global Health* 8 (8): e1003–17. https://doi.org/10.1016/S2214-109X(20)30264-3.

COG-UK. 2020. "An Integrated National Scale SARS-CoV-2 Genomic Surveillance Network." *The Lancet. Microbe* 1 (3): e99–100. https://doi.org/10.1016/S2666-5247(20)30054-9.

Cunningham, Anne Catherine, Hui Poh Goh, and David Koh. 2020. "Treatment of COVID-19: Old Tricks for New Challenges." *Critical Care* 24 (1): 91. https://doi.org/10.1186/s13054-020-2818-6.

Cyranoski, David. 2021. "Alarming COVID Variants Show Vital Role of Genomic Surveillance." *Nature* 589 (7842). Nature Publishing Group: 337–38. https://doi.org/10.1038/d41586-021-00065-4.

Deng, Xianding, Wei Gu, Scot Federman, Louis du Plessis, Oliver G. Pybus, Nuno R. Faria, Candace Wang, et al. 2020. "Genomic Surveillance Reveals Multiple Introductions of SARS-CoV-2 into Northern California." *Science* 369 (6503). American Association for the Advancement of Science: 582–87. https://doi.org/10.1126/science.abb9263.

Dong, Mengzhen, Jie Zhang, Xuefeng Ma, Jie Tan, Lizhen Chen, Shousheng Liu, Yongning Xin, and Likun Zhuang. 2020. "ACE2, TMPRSS2 Distribution and Extrapulmonary Organ Injury in Patients with COVID-19." *Biomedicine & Pharmacotherapy* 131 (November): 110678. https://doi.org/10.1016/j.biopha.2020.110678.

Furuta, Yousuke, Takashi Komeno, and Takaaki Nakamura. 2017. "Favipiravir (T-705), a Broad Spectrum Inhibitor of Viral RNA Polymerase." *Proceedings of the Japan Academy. Series B, Physical and Biological Sciences* 93 (7): 449–63. https://doi.org/10.2183/pjab.93.027.

Gao, Zhiru, Yinghui Xu, Chao Sun, Xu Wang, Ye Guo, Shi Qiu, and Kewei Ma. 2021. "A Systematic Review of Asymptomatic Infections with COVID-19." *Journal of Microbiology, Immunology and Infection* 54 (1): 12–16. https://doi.org/10.1016/j.jmii.2020.05.001.

Guan, Wei-jie, Zheng-yi Ni, Yu Hu, Wen-hua Liang, Chun-quan Ou, Jian-xing He, Lei Liu, et al. 2020. "Clinical Characteristics of Coronavirus Disease 2019 in China." *New England Journal of Medicine*, February. Massachusetts Medical Society. https://doi.org/10.1056/NEJMoa2002032.

"Guidance for COVID-19 | CDC." 2021. Accessed March 25. www.cdc.gov/coronavirus/2019-ncov/communication/guidance.html.

Hemrajata, Peera. 2021. "SARS-CoV-2 Sequencing Data: The Devil Is in the Genomic Detail." *ASM.Org*. Accessed May 5. https://asm.org/Articles/2020/October/SARS-CoV-2-Sequencing-Data-The-Devil-Is-in-the-Gen.

Hoffmann, Markus, Heike Hofmann-Winkler, Joan C. Smith, Nadine Krüger, Prerna Arora, Lambert K. Sørensen, Ole S. Søgaard, et al. 2021. "Camostat Mesylate Inhibits SARS-CoV-2 Activation by TMPRSS2-Related Proteases and Its Metabolite GBPA Exerts Antiviral Activity." *EbioMedicine* 65 (March). Elsevier. https://doi.org/10.1016/j.ebiom.2021.103255.

Hu, Zhiliang, Ci Song, Chuanjun Xu, Guangfu Jin, Yaling Chen, Xin Xu, Hongxia Ma, et al. 2020. "Clinical Characteristics of 24 Asymptomatic Infections with COVID-19 Screened among Close Contacts in Nanjing, China." *Science China. Life Sciences* 63 (5): 706–11. https://doi.org/10.1007/s11427-020-1661-4.

Huang, Chaolin, Yeming Wang, Xingwang Li, Lili Ren, Jianping Zhao, Yi Hu, Li Zhang, et al. 2020. "Clinical Features of Patients Infected with 2019 Novel Coronavirus in Wuhan, China." *The Lancet* 395 (10223). Elsevier: 497–506. https://doi.org/10.1016/S0140-6736(20)30183-5.

Kronbichler, Andreas, Daniela Kresse, Sojung Yoon, Keum Hwa Lee, Maria Effenberger, and Jae Il Shin. 2020. "Asymptomatic Patients as a Source of COVID-19 Infections: A Systematic Review and Meta-Analysis." *International Journal of Infectious Diseases* 98 (September): 180–86. https://doi.org/10.1016/j.ijid.2020.06.052.

Kucirka, Lauren M., Stephen A. Lauer, Oliver Laeyendecker, Denali Boon, and Justin Lessler. 2020. "Variation in False-Negative Rate of Reverse Transcriptase Polymerase Chain Reaction—Based SARS-CoV-2 Tests by Time Since Exposure." *Annals of Internal Medicine*, May. https://doi.org/10.7326/M20-1495.

Li, Yuchong, Mirjana Jerkic, Arthur S. Slutsky, and Haibo Zhang. 2020. "Molecular Mechanisms of Sex Bias Differences in COVID-19 Mortality." *Critical Care* 24 (1): 405. https://doi.org/10.1186/s13054-020-03118-8.

Lim, Soo, Jae Hyun Bae, Hyuk-Sang Kwon, and Michael A. Nauck. 2021. "COVID-19 and Diabetes Mellitus: From Pathophysiology to Clinical Management." *Nature Reviews Endocrinology* 17 (1). Nature Publishing Group: 11–30. https://doi.org/10.1038/s41574-020-00435-4.

Mesoraca, Alvaro, Katia Margiotti, Antonella Viola, Antonella Cima, Davide Sparacino, and Claudio Giorlandino. 2020. "Evaluation of SARS-CoV-2 Viral RNA in Fecal Samples." *Virology Journal* 17 (1): 86. https://doi.org/10.1186/s12985-020-01359-1.

Messner, Christoph B., Vadim Demichev, Daniel Wendisch, Laura Michalick, Matthew White, Anja Freiwald, Kathrin Textoris-Taube, Spyros I. Vernardis, Anna-Sophia Egger, and Marco Kreidl. 2020. "Ultra-High-Throughput Clinical Proteomics Reveals Classifiers of COVID-19 Infection." *Cell Systems* 11 (1). Elsevier: 11–24.

Naqvi, Ahmad Abu Turab, Kisa Fatima, Taj Mohammad, Urooj Fatima, Indrakant K. Singh, Archana Singh, Shaikh Muhammad Atif, Gururao Hariprasad, Gulam Mustafa Hasan, and Md. Imtaiyaz Hassan. 2020. "Insights into SARS-CoV-2 Genome, Structure, Evolution, Pathogenesis and Therapies: Structural Genomics Approach." *Biochimica et Biophysica Acta (BBA)—Molecular Basis of Disease* 1866 (10): 165878. https://doi.org/10.1016/j.bbadis.2020.165878.

Novosad, Paul, Radhika Jain, Alison Campion, and Sam Asher. 2020. "COVID-19 Mortality Effects of Underlying Health Conditions in India: A Modelling Study." *BMJ Open* 10 (12): e043165. https://doi.org/10.1136/bmjopen-2020-043165.

Nutho, Bodee, Panupong Mahalapbutr, Kowit Hengphasatporn, Nawanwat Chainuwong Pattaranggoon, Nattapon Simanon, Yasuteru Shigeta, Supot Hannongbua, and Thanyada Rungrotmongkol. 2020. "Why Are Lopinavir and Ritonavir Effective against the Newly Emerged Coronavirus 2019? Atomistic Insights into the Inhibitory Mechanisms." *Biochemistry*, April. https://doi.org/10.1021/acs.biochem.0c00160.

Oran, Daniel P., and Eric J. Topol. 2020. "Prevalence of Asymptomatic SARS-CoV-2 Infection." *Annals of Internal Medicine* 173 (5). American College of Physicians: 362–67. https://doi.org/10.7326/M20-3012.

Oude Munnink, Bas B., David F. Nieuwenhuijse, Mart Stein, Áine O'Toole, Manon Haverkate, Madelief Mollers, Sandra K. Kamga, et al. 2020. "Rapid SARS-CoV-2 Whole-Genome Sequencing and Analysis for Informed Public Health Decision-Making in the Netherlands." *Nature Medicine* 26 (9): 1405–10. https://doi.org/10.1038/s41591-020-0997-y.

Ritchie, Hannah, Edouard Mathieu, Lucas Rodés-Guirao, Cameron Appel, Charlie Giattino, Esteban Ortiz-Ospina, Joe Hasell, Bobbie Macdonald, Diana Beltekian, and Max Roser. 2020. "Coronavirus Pandemic (COVID-19)." *Our World in Data*, March. https://ourworldindata.org/coronavirus.

Rubino, Francesco, Stephanie A. Amiel, Paul Zimmet, George Alberti, Stefan Bornstein, Robert H. Eckel, Geltrude Mingrone, et al. 2020. "New-Onset Diabetes in Covid-19." *New England Journal of Medicine* 383 (8). Massachusetts Medical Society: 789–90. https://doi.org/10.1056/NEJMc2018688.

Scohy, Anaïs, Ahalieyah Anantharajah, Monique Bodéus, Benoît Kabamba-Mukadi, Alexia Verroken, and Hector Rodriguez-Villalobos. 2020. "Low Performance of Rapid Antigen Detection Test as Frontline Testing for COVID-19 Diagnosis." *Journal of Clinical Virology* 129 (August): 104455. https://doi.org/10.1016/j.jcv.2020.104455.

Sekizuka, Tsuyoshi, Kentaro Itokawa, Koji Yatsu, Rina Tanaka, Masanori Hashino, Tetsuro Kawano-Sugaya, Makoto Ohnishi, Takaji Wakita, Makoto Kuroda, and COVID-19 Genomic Surveillance Network in Japanese Airport Quarantine. 2021. "COVID-19 Genome Surveillance at International Airport Quarantine Stations in Japan." *Journal of Travel Medicine* 28 (2). https://doi.org/10.1093/jtm/taaa217.

Shen, Bo, Xiao Yi, Yaoting Sun, Xiaojie Bi, Juping Du, Chao Zhang, Sheng Quan, et al. 2020. "Proteomic and Metabolomic Characterization of COVID-19 Patient Sera." *Cell* 182 (1): 59–72.e15. https://doi.org/10.1016/j.cell.2020.05.032.

Song, Jin-Wen, Sin Man Lam, Xing Fan, Wen-Jing Cao, Si-Yu Wang, He Tian, Gek Huey Chua, et al. 2020. "Omics-Driven Systems Interrogation of Metabolic Dysregulation in COVID-19 Pathogenesis." *Cell Metabolism* 32 (2): 188–202.e5. https://doi.org/10.1016/j.cmet.2020.06.016.

Su, Yapeng, Daniel Chen, Dan Yuan, Christopher Lausted, Jongchan Choi, Chengzhen L. Dai, Valentin Voillet, et al. 2020. "Multi-Omics Resolves a Sharp Disease-State Shift between Mild and Moderate COVID-19." *Cell* 183 (6): 1479–95.e20. https://doi.org/10.1016/j.cell.2020.10.037.

Su, Shuo, Gary Wong, Weifeng Shi, Jun Liu, Alexander C. K. Lai, Jiyong Zhou, Wenjun Liu, Yuhai Bi, and George F. Gao. 2016. "Epidemiology, Genetic Recombination, and Pathogenesis of Coronaviruses." *Trends in Microbiology* 24 (6): 490–502. https://doi.org/10.1016/j.tim.2016.03.003

Suvarna, Kruthi, Deeptarup Biswas, Medha Gayathri Pai, Arup Acharjee, Renuka Bankar, Viswanthram Palanivel, Akanksha Salkar, et al. 2021. "Proteomics and Machine Learning Approaches Reveal a Set of Prognostic Markers for COVID-19 Severity with Drug Re-Purposing Potential." *Frontiers in Physiology* 12. Frontiers. https://doi.org/10.3389/fphys.2021.652799.

Tai, Wanbo, Lei He, Xiujuan Zhang, Jing Pu, Denis Voronin, Shibo Jiang, Yusen Zhou, and Lanying Du. 2020. "Characterization of the Receptor-Binding Domain (RBD) of 2019 Novel Coronavirus: Implication for Development of RBD Protein as a Viral Attachment Inhibitor and Vaccine." *Cellular & Molecular Immunology* 17 (6). Nature Publishing Group: 613–20. https://doi.org/10.1038/s41423-020-0400-4.

Wang, Dawei, Bo Hu, Chang Hu, Fangfang Zhu, Xing Liu, Jing Zhang, Binbin Wang, et al. 2020. "Clinical Characteristics of 138 Hospitalized Patients with 2019 Novel Coronavirus–Infected Pneumonia in Wuhan, China." *JAMA* 323 (11). American Medical Association: 1061–69. https://doi.org/10.1001/jama.2020.1585.

"WHO Coronavirus (COVID-19) Dashboard." 2021. Accessed September 23. https://covid19.who.int.

Wolff, Dominik, Sarah Nee, Natalie Sandy Hickey, and Michael Marschollek. 2020. "Risk Factors for Covid-19 Severity and Fatality: A Structured Literature Review." *Infection*, August: 1–14. https://doi.org/10.1007/s15010-020-01509-1.

Wu, Fan, Su Zhao, Bin Yu, Yan-Mei Chen, Wen Wang, Zhi-Gang Song, Yi Hu, et al. 2020. "A New Coronavirus Associated with Human Respiratory Disease in China." *Nature* 579 (7798). Nature Publishing Group: 265–69. https://doi.org/10.1038/s41586-020-2008-3.

Yadav, Radha. 2021. "Mumbai Mayhem of COVID-19 Pandemic Reveals Important Factors That Influence Susceptibility to Infection." *Eclinical Medicine* 35. Elsevier.

Zaim, Sevim, Jun Heng Chong, Vissagan Sankaranarayanan, and Amer Harky. 2020. "COVID-19 and Multiorgan Response." *Current Problems in Cardiology* 45 (8): 100618. https://doi.org/10.1016/j.cpcardiol.2020.100618.

Zhang, Xiuhong, Yan Zhang, Weizhen Qiao, Ji Zhang, and Zhigang Qi. 2020. "Baricitinib, a Drug with Potential Effect to Prevent SARS-COV-2 from Entering Target Cells and Control Cytokine Storm Induced by COVID-19." *International Immunopharmacology* 86 (September): 106749. https://doi.org/10.1016/j.intimp.2020.106749.

Zhu, Na, Dingyu Zhang, Wenling Wang, Xingwang Li, Bo Yang, Jingdong Song, Xiang Zhao, Baoying Huang, Weifeng Shi, and Roujian Lu. 2020. "A Novel Coronavirus from Patients with Pneumonia in China, 2019." *New England Journal of Medicine* 382 (8). Massachusetts Medical Society: 727–33.

2

NGS technologies for detection of SARS-CoV-2 strains and mutations

Manisha Choudhury, Ayushi Verma, Ankit Halder, and Arup Acharjee

CONTENTS

DOI: 10.1201/9781003220787-2

List of abbreviations

BAM – Binary alignment map
Ct – Threshold cycle
GISRS – Global influenza surveillance and response system
gRNA – Genomic RNA
MERS – CoV-Middle East respiratory syndrome coronavirus
NGS – Next-generation sequencing
NFW – Nuclease-free water
Nsps – Non-structural proteins
ORFs – Open reading frames
RdRP – RNA-dependent RNA polymerase
RT-PCRs – Reverse-transcriptase polymerase chain reactions
sgRNAs – Subgenomic RNAs
SNP – Single-nucleotide polymorphisms
VCF – Variant call format
WGS – Whole-genome sequencing
WHO – World Health Organization

2.1 Importance of SARS-CoV-2 whole-genome sequencing

One of the most powerful technologies that provide a high-resolution view of the pathogen evolution and helps in analyzing the transmission dynamics of outbreaks is whole-genome sequencing (WGS). The significant achievements of WGS in COVID-19 are the designing of specific reverse-transcriptase polymerase chain reactions (RT-PCRs) (Corman et al. 2020), antiviral drug candidates (Dai et al. 2020), and vaccine candidates (Kames et al. 2020). The involvement of WGS in public health decision-making was observed during the West African Ebola outbreak in 2014, and the strategy was termed "precision public health pathogen genomics" (Quick et al. 2016; Gire et al. 2014). The practical use of WGS for outbreak support is possible subjective to the availability of reliable sequencing platforms, accessible metadata, and faster data analysis. During recent outbreaks of Usutu, Ebola, Zika, and yellow fever virus, WGS using nanopore technology have already been reported (Oude Munnink et al. 2020; Quick et al. 2016; Gire et al. 2014; Nuno Rodrigues Faria et al. 2016; N. R. Faria et al. 2018). Several studies have shown that the WGS of SARS-CoV-2 has enabled us to understand the virus biology, transmission, and population dynamics during the current pandemic. Genomic surveillance of SARS-CoV-2 can reveal the prevalent virus lineage in the population, mutations, the effect of control measures, and the virus's evolution due to interventions.

2.2 Understanding the recent SARS-CoV-2 mutations using next-generation sequencing

With the pandemic outbreak in China in late 2019, sequencing technologies could help identify the causative organism of the pandemic, i.e., SARS-CoV-2, a type of betacoronavirus. The Coronaviridae Study Group (CSG) of the International Committee on Taxonomy of Viruses placed the pathogen within the Coronaviridae and tentatively named it 2019-nCoV. Phylogenetic analysis revealed that the virus formed a sister clade with human and bat SARS-CoVs of the severe acute respiratory syndrome-related coronaviruses. Therefore, it was referred to as SARS-CoV-2 (Coronaviridae Study Group of the International Committee on Taxonomy of Viruses 2020).

The preliminary investigation also revealed the genomic structure and organization of SARS-CoV-2. Like any other *Betacoronaviruses*, it comprises ORF1ab polyprotein at 5′-end and four major structural proteins – the spike surface glycoprotein, small envelope protein, matrix protein, and nucleocapsid protein. Further, the coronaviral RNA remains encapsulated by a 5′ "leader" sequence of approximately 70 nucleotides long. The "body" sequence remains in continuity with the "leader" sequence and makes the downstream part of the genome (Phan 2020a).

One of the essential characteristics of the COVID-19 pandemic has been the events of superspreading. Using WGS data, Lemieux et al. could identify superspreading events and analyze them phylogenetically (Lemieux et al. 2021). The genomic sequence analysis led Summer E. Galloway et al. to identify the B.1.1.7 lineage in the United States, which the United Kingdom reported on December 14, 2020 (Galloway et al. 2021). Similarly, after sequencing 180 complete genomes in Brazil, Carolina M. Voloch et al. could identify 37 samples comprising a new variant lineage. The report stated the new lineage to be a descendant of the B.1.1.28 strain and comprises five striking mutations. The mutations were 5′ UTR C100U, ORF1ab L3468V, ORF1ab 34 synC11824U, S E484K, and N A119S. They identified 731 SNVs across all samples and reported missense, synonymous, intergenic, and nonsense mutations (Voloch et al. 2021). After the reports of the new variants in the UK (20I/501Y.V1/B.1.1.7), South Africa (20H/501Y.V2/B.1.351), and Brazil (P. 1/20J/501Y.V3/B.1.1.248), Wenjuan Zhang et al. analyzed variants in Southern California. On performing phylogenetic analysis, they could identify lineages with two main clusters. They observed that CAL.20C lineage was first reported in October 2020 (W. Zhang et al. 2021). Nextstrain estimates the rate of mutation at around 23bases/year (Figure 2.1).

2.3 The SARS-CoV-2 genome structure

In the family Coronaviridae to which coronaviruses belong, the positive-sense single-stranded RNA makes up the genome. Their genome size ranges from 25 to 32 kb, and they are one of the largest known RNA viruses. The family Coronaviridae placed under the order Nidovirales is further subdivided into two subfamilies: Coronavirinae and the Torovirinae. The type of nucleocapsids they possess is the basis of the subdivision of the family. Further, the subfamily Coronavirinae is divided into four genera: the alpha-, beta-, gamma-, and delta coronaviruses that are extensively distributed within mammals (Payne 2017).

SARS-CoV-2 belongs to the genus *Betacoronaviruses*, comprised of a positive-sense, single-stranded RNA genome of around 30 kb and shares 80% homology SARS-CoV and 50% homology with Middle East respiratory syndrome coronavirus

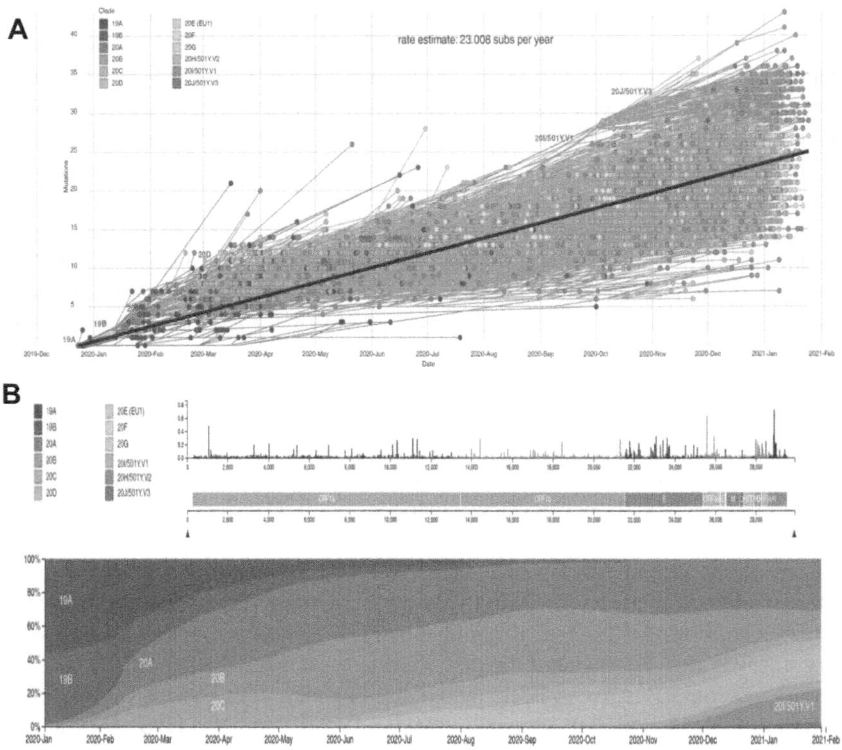

FIGURE 2.1 The real-time tracking of SARS-CoV-2 mutations using Nextstrain https://nextstrain.org/. The estimated rate of mutations per year and the genome region with a higher rate of mutations are shown.

(MERS-CoV). The presence of a 5′-cap structure and a 3′ poly(A) tail characterizes the transcript of the coronavirus (COVs). The translation of two open reading frames (ORFs), namely, ORF1a and ORF1b, produces non-structural proteins (nsps) (Figure 2.2). Polypeptide 1a (pp1a) of molecular weight 440–500 kDa is made from ORF1a, which is further cleaved into 11 nsps. The translation of ORF1b produces a polypeptide (pp1ab) of molecular weight ranging around 740–810 kDa of protein. This is cleaved into 15 nsps. Among these, nsp3 and nsp5 are the two main proteases responsible for proteolytic cleavage. Nsp3 bears a papain-like protease domain, whereas nsp5 contains a 3C-like protease domain. The nsp12 mediates the viral genome replication and transcription by catalyzing RNA-dependent RNA polymerase (RdRP) activity.

The process of replication involves generating negative-sense RNA intermediates. These intermediates work as templates to synthesize positive-sense genomic RNA (gRNA) and subgenomic RNAs (sgRNAs). The gRNA is responsible for encoding the proteins that assemble progeny virions, whereas sgRNA encodes conserved structural proteins like spike protein (S), envelope protein (E), membrane protein (M), and nucleocapsid protein (N) and several other accessory proteins. There are at least

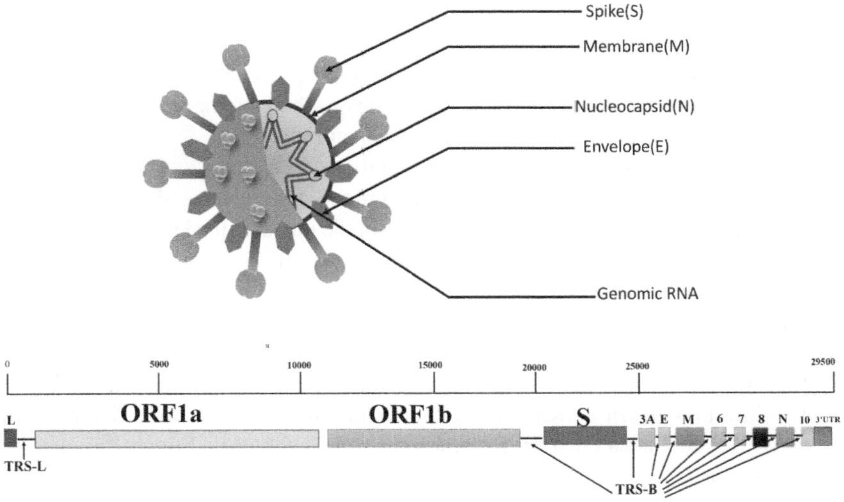

FIGURE 2.2 Schematic representation of the SARS-CoV-2 structure. Representation of the SARS-CoV-2 genome showing the protein-coding regions.

six types of accessory proteins (3a, 6, 7a, 7b, 8, and 10) known in SARS-CoV-2. The coronaviral RNA remains encapsulated by a 5′ "leader" sequence of approximately 70 nucleotides long. The "body" sequence remains in continuity with the "leader" sequence and makes the downstream part of the genome. The general mechanism of replication of SARS-CoV involves transcription-regulatory sequences (TRSs). These are short motifs found adjacent to the ORFs and are the site for the fusion of "leader" and "body" sequences during negative-strand synthesis. Discontinuous transcription is a phenomenon that is prevalent in COVs. It is caused due to pausing of RdRp whenever it crosses a TRS in the body (TRS-B), it switches the template to the TRS in the leader (TRS-L), and leads to the generation of a leader–body fusion. About 41 RNA modification sites were detected in the transcript with a shorter poly(A) tail, and the most common motif identified was AAGAA (Kim et al. 2020). In addition to the canonical RNAs (both gRNAs and sgRNAs), SARS-CoV-2 contains additional transcripts that encode unknown ORFs with fusion, deletion, and/or frameshift.

2.4 SARS-CoV-2 genome in terms of genetic diversity and evolution

The studies on the genome of SARS-CoV-2 reveal that, like any other *Betacoronaviruses*, it comprises ORF1ab polyprotein at 5′-end and four major structural proteins, the spike surface glycoprotein, small envelope protein, matrix protein, and nucleocapsid protein (Phan 2020b). In terms of mutations, one of the studies revealed three deletions in the genomes of SARS-CoV-2; two deletions (comprising 3 and 24 nucleotides) were found in the ORF1ab polyprotein, and one deletion of 10 nucleotides was noticed in the 3′-end of the genome in the SARS-CoV-2 samples from different countries. The other significant finding was that overall 93 mutations could be detected in the entire genome of SARS-CoV-2. The major nsp and structural protein-coding regions in the genome incurred missense mutations, the only exception being the envelope protein.

Further annotating the mutations, 29 mutations were reported in the ORF1ab polyprotein, eight types of missense mutation in the spike surface glycoprotein, one was discerned in the matrix protein, and four in the nucleocapsid protein. The most notable three findings could be linked with the report of mutations in the D354, Y364, and F367 located in the spike-surface glycoprotein and the receptor-binding domain. The mutations resulting in the conformational change of the glycoproteins also differ in antigenicity (Phan 2020b).

From the widespread infections of SARS-CoV-2, it is evident that it transmits to and within humans. In SARS-CoV-2, its genome shared nearly 96% identity with BatCoV RaTG13 isolated from *Rhinolophus affinis* found in Yunan Province in China. It was crucial to identify intermediate hosts, if any, to block the interspecies transmission. At almost the same time the outbreak occurred, SARS-CoV-like CoV from dead Malayan pangolins was extracted. The viral genome isolated from pangolin shared nearly 91.02% and 90.55% identical regions with the SARS-CoV-2 and BatCoV RaTG13 genomes, respectively. The S1 protein of Pangolin-CoV further supported this finding. Pangolin-CoV and SARS-CoV-2 shared the exact five amino acid sequences required to interact with human ACE2, but a slight variation of amino acids is present in RaTG13. The S1/S2 cleavage site remains intact in all three CoVs. Thus, it was concluded that the pangolin species possess reserves of SARS-CoV-2-like viruses (T. Zhang, Wu, and Zhang 2020) and are a probable intermediate species responsible for the outbreak.

2.5 Global collaborations to collect and analyze SARS-CoV-2 sequence data

The World Health Organization's (WHO's) Global Influenza Surveillance Network (now the Global Influenza Surveillance and Response System [GISRS]) was set up by Los Alamos National Laboratories in the United States to share sequencing data on the H5N1 influenza virus. Scientists from WHO Collaborating Centres (CCs) for Influenza, National Influenza Centres, and the World Organisation for Animal Health (OIE)/Food and Agriculture Organization of the United Nations (FAO) reference laboratories came up with the idea of creation and development of "*A global initiative on sharing avian flu data*" (GISAID) platform to share sequencing data. The initiative resulted in the creation of the GISAID Initiative in 2008. The intention was to promote sharing of all influenza virus data and metadata among the community worldwide. The GISAID Database Access Agreement (DAA) was enforced to decide on the rules for sharing genetic data to meet emergencies, including intellectual property rights. With its launch of EpiFlu Database in 2008, the importance was understood during the 2009 influenza A (H1N1) outbreak that helped the researchers to follow the evolutionary pattern of the virus that spread globally. And, now in the case of COVID-19 outbreaks, its importance is unprecedented, helping researchers to track epidemiology, the variants, evolution, and most importantly in deciding the candidate vaccine targets (Delaunay et al. 2016).

2.6 SARS-CoV-2 whole-genome sequencing workflow

With the advances of next-generation sequencing, high-throughput sequencing can be performed at reduced time and cost. The most popular high-throughput NGS instruments are Illumina platforms, Ion Torrent S5 (Thermo Fisher), Roche 454, and Oxford Nanopore. Mainly all NGS platforms utilize a similar workflow, including library/template preparation, template amplification, sequencing, and data analysis (Figure 2.3).

2.7 Steps involved in the SARS-CoV-2 whole-genome sequencing

2.7.1 Sample collection and preparation

For SARS-CoV-2 whole-genome sequencing, clinical specimens, mostly nasopharyngeal or oropharyngeal swabs from RT-PCR-confirmed COVID-19 patients, are collected. RNA from the samples is extracted using Trizol or RNA extraction kits like QIAamp Viral RNA Mini Kit (Qiagen, Valencia, CA). RNA samples are quantified using Qubit RNA Assay HS kit (Invitrogen) and stored at −80°C before use.

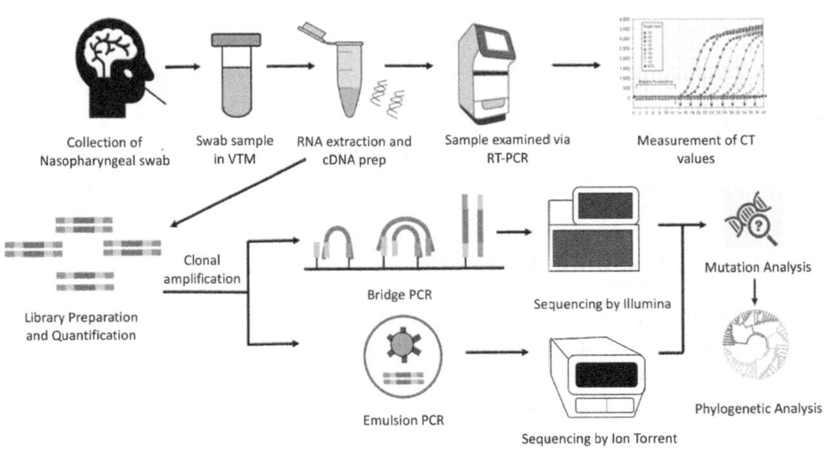

FIGURE 2.3 Overview of the next-generation sequencing workflow for SARS-CoV-2 whole-genome sequencing using Illumina and Ion Torrent platforms.

2.7.2 Real-time polymerase chain reaction

For RT-PCR testing, RNA samples are thawed at room temperature. Three SARS-CoV-2 genes—ORF1ab, S protein, and N protein—are tested using the TaqMan 2019-nCoV Assay Kit (LifeTechnologies, USA) to detect SARS-CoV-2 infection. A master mix is prepared for each target gene which includes nuclease-free water (NFW), TaqMan Master Mix, Rnase P, and 2019-nCoV target. A positive control containing TaqMan 2019-nCoV control kit and NFW and no template control containing NFW for each target gene is included. RT-PCR is performed using the following conditions: 50°C for 5 minutes, 95°C for 20 seconds, 40 cycles at 95°C for 3 seconds, and 60°C for 30 seconds. Ct values of each sample are analyzed, and the Ct value of RNAse P is used as an internal amplification control.

Ct value is the number of RT-PCR cycles required to amplify the viral RNA to exceed the threshold value and reach the detectable concentration. In general, it has been observed that the lower the Ct value of a sample, the higher the viral load in the sample and vice versa ("Real-Time PCR: Understanding Ct|Thermo Fisher Scientific—IN" n.d.). Samples with the same Ct values are taken forward for cDNA synthesis as the samples with high template concentration can interfere in amplifying the samples with lower template concentration.

2.7.3 cDNA synthesis

The complementary DNA is synthesized using SuperScript™ VILO™ cDNA Synthesis Kit (Thermo Fisher Scientific) containing 5X VILO™ Reaction Mix and 10X SuperScript™ III Enzyme Mix. First random hexamer primers are used, followed by gene-specific primers using the ARTIC protocol (Quick, J. nCoV-2019 sequencing protocol V.1.). After the cDNA synthesis, PCR products are purified using AmpureXP purification beads. Quantification is performed using Qubit double-strand DNA (dsDNA) High Sensitivity assay kit (Thermo Fisher Scientific) on a Qubit fluorometer.

2.7.4 Library preparation

The first step in sequencing involves library preparation. It includes fragmentation of cDNA into smaller amplicons and ligation of oligonucleotide adapters to the ends of the amplicons. The prepared library is then purified and quantified for the next step.

(i) Library preparation for the Illumina AmpliSeq

The AmpliSeq Library Plus kits by Illumina (Illumina, San Diego, CA) involve preparing paired-end libraries. The Focus Panels include an integrated workflow that comprises AmpliSeq for Illumina polymerase chain reaction (PCR)-based library preparation, Illumina sequencing by synthesis (SBS) NGS technology, and automated analysis. The library is quantified using a Qubit dsDNA High Sensitivity assay kit on a Qubit fluorometer (Thermo Fisher Scientific). Further, the library is spiked with 1% PhiX control, and sequencing is performed.

(ii) Ion AmpliSeq library preparation

Sequencing on the Ion Torrent platform includes the library preparation using Ion AmpliSeq™ Library Kit Plus (Thermo Fisher Scientific). 5X Ion AmpliSeq™ HiFi Mix is mixed with the Ion AmpliSeq ™ panels and NFW. For 100 ng of RNA, ten cycles of thermocycler are performed. Adapter ligation is performed using Ion P1 Adapters, and samples are annotated using Ion Xpress™ Barcode X. Further, the amplified library is purified using AMPure™ XP Reagent and quantified using Agilent™ 2100 Bioanalyzer™ Instrument.

2.7.5 Clonal amplification

In the Ion Torrent platform (Thermo Fisher), emulsion PCR is used for clonal amplification of adapter-ligated nucleic acid fragments on the surface of the beads in oil and aqueous mixture. Illumina platform utilizes the isothermal bridge amplification of adaptor-ligated RNA fragments on the surface of the glass slide.

2.7.6 Sequencing

The amplified library, now called enriched ion sphere particles, is evenly distributed on a semiconductor chip with millions of microwells for sequencing in the Ion Torrent platform. During DNA extension, when the phosphodiester bond is formed, a hydrogen ion is released. This hydrogen ion or change in pH is detected by a sensor placed at the bottom of the microwell, which converts this signal into a voltage signal. The voltage signal indicates the addition of a new nucleotide in the sequencing strand and is directly proportional to the number of bases incorporated. Illumina sequencing is performed on a flow cell, where complementary fluorescently labeled 3′-*O*-azidomethyldNTPs are added, giving the signal for sequencing. The process is called sequencing by synthesis.

2.7.7 Data analysis

The WGS data reveals insights about the genomic variants like single-nucleotide polymorphisms (SNPs), insertions, and deletions that drive the organism toward new strains to counter the host's immune system. For instance, the phenomenon of "Antigenic Drift" and "Antigenic Shift" is quite common in influenza viruses (H. Kim, Webster, and Webby 2018). It is noteworthy that a new lineage, "B.1.1.7", could be detected, and several significant mutations could be identified. The rapid extension of the lineages signifies the importance of genomic and epidemiological surveillance ("Preliminary Genomic Characterisation of an Emergent SARS-CoV-2 Lineage in the UK Defined by a Novel Set of Spike Mutations—SARS-CoV-2 Coronavirus/NCoV-2019 Genomic Epidemiology" 2020).

The data analysis pipeline of WGS involves numerous critical steps, from the quality check of input data to the detection of genomic variants like SNPs or insertions and deletions (Figure 2.4). The steps that are generally followed are mentioned as follows.

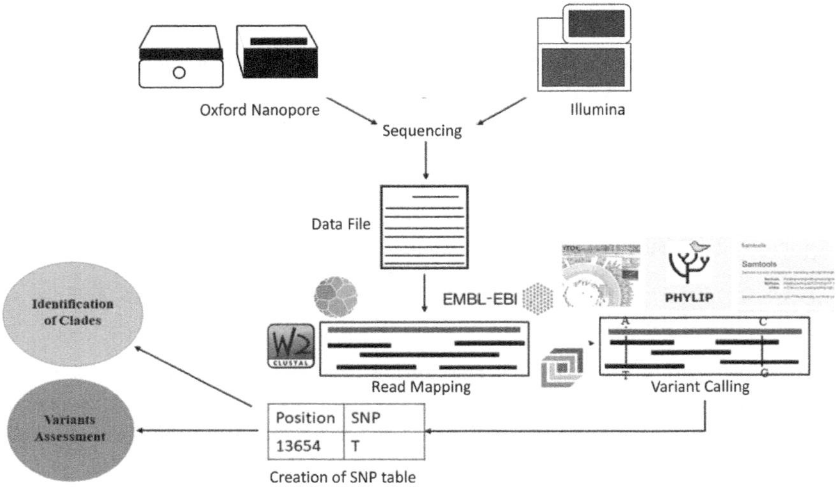

FIGURE 2.4 Schematic of the data analysis workflow. Analysis of the SARS-CoV-2 genome sequencing data obtained using Illumina or Oxford Nanopore platforms.

(i) Quality control and preprocessing of the sequence data

Quality check of the FASTQ files (RNA sequencing data) must be carried out before further downstream analyses to remove the adaptors or poor-quality bases. Software like FastQC (www.bioinformatics.babhraham.ac.uk/projects/fastqc/) can be employed to assess the quality of the fastq files generated. To remove these poor-quality bases, tools like TrimGalore (www.bioinformatics.babraham.ac.uk/projects/trim_galore/) or Trimmomatic (www.useadellab.org/cms/?page=trimmomatic) can be used (Banu et al. 2020). Some other criteria that can filter out the sequences are the number of Ns or gaps greater than 1%.

(ii) Alignment to reference genome

Sequence alignment is the next process to identify the variants or cluster similar sequences to identify the virus's different clades. In the analysis of the sequences of SARS-CoV-2, the reference genome is generally taken as Wuhan/WHO1/2019 (EPI_ISL_406798) (NC_045122), the first strain reported. Reads can be aligned to the reference genome using a tool like hisat2 (Banu et al. 2020). For the construction of the phylogenetic tree, generally, multiple sequence alignment (MSA) is performed using tools like MAFFT (https://mafft.cbrc.jp/alignment/software/), MUSCLE (www.ebi.ac.uk/Tools/msa/muscle/), T-Coffee (www.ebi.ac.uk/Tools/msa/tcoffee), and ClustalW2 (www.clustal.org/clustal2/). Nextstrain (https://nextstrain.org/), the software developed by Hadfield et al., provides a phylogenetic analysis pipeline. It makes use of Augur to assemble similar sequences. The MSA is performed using MAFFT to create an alignment file. The next step toward the identification of variants is the generation of the consensus sequence. The quality check should be performed on the binary alignment map (BAM) file generated to filter out the low-quality score and bcftools (http://samtools.github.io/bcftools/bcftools.html) can be employed to manipulate variant calls in the Variant Call Format (VCF).

(iii) Construction of phylogenetic tree and identification of the ancestral traits

The MSA files are generated in the aln format by ClustalW and can be visualized in software like Jalview (www.jalview.org/) for analysis. The pipeline provided by Nextstrain uses IQTREE after the initial cluster formed by Augur. Another popular software is PHYLIP (https://evolution.genetics.washigton.edu/phylip.html) used for inferring phylogeny. A study used BEAST v1.10.4 for the understanding of nucleotide substitution rates. It also estimated the time to the most recent common ancestor (Banu et al. 2020). The BEAST performs Bayesian coalescent analysis to calculate the times to the most common ancestor for the individual clusters. The trees generally generated in NEWICK tree format can be interactively visualized in webservers like iTOL (Interactive Tree of Life) (https://itol.embl.de/). It provides a platform to differentiate the different clades.

2.7.8 Assessing the impacts of the variants identified

The variants that distinguish the virus strains from a particular clade need to be assessed for the impact on the overall organization of the virus's genome. The Protein Variation Effect Analyzer (PROVEAN) can be found suitable for predicting the effect of amino acid substitution or indel on the biological function of a protein. Another tool, the Sorting Intolerant from Tolerant (SIFT), is used to predict the deleterious effects of the mutated protein. The variations in the epitope regions or vaccine targets can be a potential way of survival for the virus by evading the host immune mechanism. For instance, Lucy van Dorp et al. reported how the genome sequence analysis reveals insights regarding the adaptation of the virus in its novel human host. They could identify regions of the genome that are conserved and the ones that are frequently mutating (van Dorp et al. 2020).

2.8 Application of whole-genome sequencing in understanding fundamental SARS-CoV-2 biology and lineage

The implications of the genotyping of SARS-CoV-2 and the need to adapt proper sequencing technologies have become quite evident. A report on the method for effective analysis of SARS-CoV-2 genomes stated to perform multiple sequence alignments of the genomes isolated from SARS-CoV-2 with a reference genome. Then Jaccard distances are to be employed so that the genotypes showing single-nucleotide polymorphism (SNP) can be tracked to identify their relationship. By deploying this strategy, it could detect that the genes encoding the S proteins and RNA polymerase, RNA primase, and nucleoprotein are the ones that are undergoing frequent mutations (Yin 2020). The phylogeographic patterns of various strains were drawn with the sequence data to identify the disease's transmission patterns that could direct the invention of diagnostics and future vaccine design. The study by Alberto Gomez-Carballa et al. concluded that the SARS-CoV-2 genome closely resembled bat CoV and pangolin CoV and subdivided it into two micro-haplogroups, A and B, with subbranches representing strains worldwide. It revealed a kind of uniform mutation across all branches (Gómez-Carballa et al. 2020). Sudhir Kumar et al. attempted to analyze the mutational history of SARS-CoV-2 genomes using the mutation order approach (MOA) that does not require phylogeny analysis as an intermediate step. It becomes evident that SARS-CoV-2 genomes follow clonal evolution, preserving the collinearity of the variants in the genome. This method could calculate the co-occurrence Index (COI) and reported around 96.9% for 49 variants that could infer the mutational history strongly. Another supporting evidence could be the emergence of μ and α variants even before the first reports of COVID-19 (Kumar et al. 2021). The study by D. Paraskevis et al. discarded any claims of the origin of SARS-CoV-2 due to recombination events and claims to identify a uniform ancestry across the genome (Paraskevis et al. 2020).

The studies focused mainly on the phylogenetic lineages and evolutionary details of SARS-CoV-2 based on the genomic sequences. However, the mutations must be analyzed for assessing the stabilizing or destabilizing impacts on critical proteins responsible for infection, virulence, and pathogenesis. For SARS-CoV-2, E, M, N, and S are such vital proteins. Because most vaccine candidates and other therapeutics target these essential proteins, it becomes crucial to know the effects of mutation on them. So, after performing phylogenetic analysis, Jobin John Jacob and his group tried to assess the impact of the mutation on the stability of the S and RBD proteins, those that are critical antigens of the vaccines. Some of the notable findings were that among the 38 amino acid substitutions they could identify in the spike protein, around 13 were stabilizing and 16 were destabilizing it. Conclusively, it was reported that the rigidity of the S protein structure increased and provided more stable conformation to the spike protein leading to more effective binding to the ACE2 receptor. The significant mutations like D614G and S477N were positioned in the potential epitope regions, while S477N is in the receptor-binding domain of the protein. However, the number of substitutions identified in M proteins was less, indicating increased stability. Therefore, it can be said that there is an outstanding balance between the stabilizing and unstabilizing mutations; and tracking it on some significant proteins and vaccine candidates becomes crucial to get notified about any kind of escape mechanism that SARS-CoV-2 would incur (Jacob et al. 2020).

2.9 COVID-19 re-infection prediction and surveillance using whole-genome sequencing data

There were few reports on secondary infection events with SARS-CoV-2 from Hong Kong, Belgium, and Ecuador (To et al. 2020; Van Elslande et al. 2020; Prado-Vivar et al. 2020). Primary illness due to SARS-CoV-2 infection followed by a distinct secondary infection with the same viral agent can be established as a reinfection after genetic analysis of the viral sequences. Tillet et al. reported genomic evidence of SARS-CoV-2 reinfection of the first North American with two distinct COVID-19 illnesses (Tillett et al. 2021). The second infection was comparatively severe, and the genomic analysis revealed that the two viral agents were genetically distinct. The significance of analyzing reinfection cases utilizing genomics approaches can be associated with vaccination response to COVID-19. Although more than 800,000 sequence submissions in GISAID have been made to date (March 22, 2021), this number lags far behind the total cases of infection detected by RT-PCR tests. This underlines the difficulty in limiting the spread of the virus and tracking its genetic sequence in real time.

Current testing procedures like RT-PCR testing can only detect the presence of the virus. No information on the virus's genetic sequence, coinfection, or host immune response can be derived using RT-PCR. With the emerging mutations, it becomes imperative to use NGS technology to support the RT-PCR technique to ensure the accuracy of the test and derive valuable information. Because of this, there has been advancement in the NGS platforms, and, currently, two methods have been devised for SARS-CoV-2 sequencing. One workflow is based on shotgun genomics and another on target enrichment. Several mutations such as amino acid deletion in nsp2 (Bal et al. 2020), truncated ORF (Batty et al. 2020), and nucleotide deletion in ORF (Holland et al. 2020) have been identified by SARS-CoV-2 genomic monitoring in the population. Early sequencing of the SARS-CoV-2 genome was carried out using viral metagenomic NGS (mNGS). However, this method lacks sensitivity in samples with low viral load, and a very high depth of sequencing is not possible (Bal et al. 2020). In such cases, targeted methods, including amplicon or capture-based enrichments, are used.

2.10 Conclusions

The complete sequence of the SARS-CoV-2 genome has been obtained using NGS, and this information has allowed the development of the PCR test, widely used to detect COVID-19 infection. Combining the diagnosis of SARS-CoV-2 infection with its genome sequence has opened a multitude of opportunities to understand its spread and evolution. The cloud-based metagenomics platforms have come up that can quickly assemble NGS data into genomes and are user-friendly with a graphical interface that anyone can use, making it fast and accessible worldwide. Studying the SARS-CoV-2 genetic makeup has been significant toward the response to the current pandemic, from understanding viral transmission pathways to tracking its evolution. A global infrastructure for genome surveillance is the utmost requirement to manage things better and more efficiently during the next pandemic, with minimum damage to our economies.

REFERENCES

Bal, A., G. Destras, A. Gaymard, M. Bouscambert-Duchamp, M. Valette, V. Escuret, E. Frobert, et al. 2020. "Molecular Characterization of SARS-CoV-2 in the First COVID-19 Cluster in France Reveals an Amino Acid Deletion in Nsp2 (Asp268del)." *Clinical Microbiology and Infection: The Official Publication of the European Society of Clinical Microbiology and Infectious Diseases* 26 (7): 960–62. https://doi.org/10.1016/j.cmi.2020.03.020.

Banu, Sofia, Bani Jolly, Payel Mukherjee, Priya Singh, Shagufta Khan, Lamuk Zaveri, Sakshi Shambhavi, et al. 2020. "A Distinct Phylogenetic Cluster of Indian Severe Acute Respiratory Syndrome Coronavirus 2 Isolates." *Open Forum Infectious Diseases* 7 (11): ofaa434. https://doi.org/10.1093/ofid/ofaa434.

Batty, Elizabeth M., Theerarat Kochakarn, Bhakbhoom Panthan, Krittikorn Kümpornsin, Poramate Jiaranai, Arporn Wangwiwatsin, Namfon Kotanan, et al. 2020. "Genomic Surveillance of SARS-CoV-2 in Thailand Reveals Mixed Imported Populations, a Local Lineage Expansion and a Virus with Truncated ORF7a." *MedRxiv*, May. https://doi.org/10.1101/2020.05.22.20108498.

Corman, Victor M., Olfert Landt, Marco Kaiser, Richard Molenkamp, Adam Meijer, Daniel K. W. Chu, Tobias Bleicker, et al. 2020. "Detection of 2019 Novel Coronavirus (2019-NcoV) by Real-Time RT-PCR." *Eurosurveillance* 25 (3): 2000045. https://doi.org/10.2807/1560-7917.ES.2020.25.3.2000045.

Coronaviridae Study Group of the International Committee on Taxonomy of Viruses. 2020. "The Species Severe Acute Respiratory Syndrome-Related Coronavirus: Classifying 2019-NcoV and Naming It SARS-CoV-2." *Nature Microbiology* 5 (4): 536–44. https://doi.org/10.1038/s41564-020-0695-z.

Dai, Wenhao, Bing Zhang, Xia-Ming Jiang, Haixia Su, Jian Li, Yao Zhao, Xiong Xie, et al. 2020. "Structure-Based Design of Antiviral Drug Candidates Targeting the SARS-CoV-2 Main Protease." *Science (New York, N.Y.)* 368 (6497): 1331–35. https://doi.org/10.1126/science.abb4489.

Delaunay, Sophie, Patricia Kahn, Mercedes Tatay, and Joanne Liu. 2016. "Knowledge Sharing during Public Health Emergencies: From Global Call to Effective Implementation." *Bulletin of the World Health Organization* 94 (4): 236–236A. https://doi.org/10.2471/BLT.16.172650.

Dorp, Lucy van, Mislav Acman, Damien Richard, Liam P. Shaw, Charlotte E. Ford, Louise Ormond, Christopher J. Owen, et al. 2020. "Emergence of Genomic Diversity and Recurrent Mutations in SARS-CoV-2." *Infection, Genetics and Evolution: Journal of Molecular Epidemiology and Evolutionary Genetics in Infectious Diseases* 83 (September): 104351. https://doi.org/10.1016/j.meegid.2020.104351.

Faria, Nuno Rodrigues, M. U. G. Kraemer, S. C. Hill, J. Goes de Jesus, R. S. Aguiar, F. C. M. Iani, J. Xavier, et al. 2018. "Genomic and Epidemiological Monitoring of Yellow Fever Virus Transmission Potential." *Science (New York, N.Y.)* 361 (6405): 894–99. https://doi.org/10.1126/science.aat7115.

Faria, Nuno Rodrigues, Raimunda do Socorro da Silva Azevedo, Moritz U. G. Kraemer, Renato Souza, Mariana Sequetin Cunha, Sarah C. Hill, Julien Thézé, et al. 2016. "Zika Virus in the Americas: Early Epidemiological and Genetic Findings." *Science (New York, N.Y.)* 352 (6283): 345–49. https://doi.org/10.1126/science.aaf5036.

Galloway, Summer E., Prabasaj Paul, Duncan R. MacCannell, Michael A. Johansson, John T. Brooks, Adam MacNeil, Rachel B. Slayton, et al. 2021. "Emergence of SARS-CoV-2 B.1.1.7 Lineage—United States, December 29, 2020–January 12, 2021." *MMWR. Morbidity and Mortality Weekly Report* 70 (3): 95–99. https://doi.org/10.15585/mmwr.mm7003e2.

Gire, Stephen K., Augustine Goba, Kristian G. Andersen, Rachel S. G. Sealfon, Daniel J. Park, Lansana Kanneh, Simbirie Jalloh, et al. 2014. "Genomic Surveillance Elucidates Ebola Virus Origin and Transmission during the 2014 Outbreak." *Science (New York, N.Y.)* 345 (6202): 1369–72. https://doi.org/10.1126/science.1259657.

Gómez-Carballa, Alberto, Xabier Bello, Jacobo Pardo-Seco, Federico Martinón-Torres, and Antonio Salas. 2020. "Mapping Genome Variation of SARS-CoV-2 Worldwide Highlights the Impact of COVID-19 Super-Spreaders." *Genome Research* 30 (10): 1434–48. https://doi.org/10.1101/gr.266221.120.

Holland, LaRinda A., Emily A. Kaelin, Rabia Maqsood, Bereket Estifanos, Lily I. Wu, Arvind Varsani, Rolf U. Halden, Brenda G. Hogue, Matthew Scotch, and Efrem S. Lim. 2020. "An 81-Nucleotide Deletion in SARS-CoV-2 ORF7a Identified from Sentinel Surveillance in Arizona (January to March 2020)." *Journal of Virology* 94 (14). https://doi.org/10.1128/JVI.00711-20.

Jacob, Jobin John, Karthick Vasudevan, Agila Kumari Pragasam, Karthik Gunasekaran, Gagandeep Kang, Balaji Veeraraghavan, and Ankur Mutreja. 2020. "Evolutionary Tracking of SARS-CoV-2 Genetic Variants Highlights Intricate Balance of Stabilizing and Destabilizing Mutations." *BioRxiv*, December. https://doi.org/10.1101/2020.12.22.423920.

Kames, Jacob, David D. Holcomb, Ofer Kimchi, Michael DiCuccio, Nobuko Hamasaki-Katagiri, Tony Wang, Anton A. Komar, Aikaterini Alexaki, and Chava Kimchi-Sarfaty. 2020. "Sequence Analysis of SARS-CoV-2 Genome Reveals Features Important for Vaccine Design." *Scientific Reports* 10 (1): 15643. https://doi.org/10.1038/s41598-020-72533-2.

Kim, Dongwan, Joo-Yeon Lee, Jeong-Sun Yang, Jun Won Kim, V. Narry Kim, and Hyeshik Chang. 2020. "The Architecture of SARS-CoV-2 Transcriptome." *Cell* 181 (4): 914–21.e10. https://doi.org/10.1016/j.cell.2020.04.011.

Kim, Hyunsuh, Robert G. Webster, and Richard J. Webby. 2018. "Influenza Virus: Dealing with a Drifting and Shifting Pathogen." *Viral Immunology* 31 (2): 174–83. https://doi.org/10.1089/vim.2017.0141.

Kumar, Sudhir, Qiqing Tao, Steven Weaver, Maxwell Sanderford, Marcos A. Caraballo-Ortiz, Sudip Sharma, Sergei L. K. Pond, and Sayaka Miura. 2021. "An Evolutionary Portrait of the Progenitor SARS-CoV-2 and Its Dominant Offshoots in COVID-19 Pandemic." *BioRxiv*, January. https://doi.org/10.1101/2020.09.24.311845.

Lemieux, Jacob E., Katherine J. Siddle, Bennett M. Shaw, Christine Loreth, Stephen F. Schaffner, Adrianne Gladden-Young, Gordon Adams, et al. 2021. "Phylogenetic Analysis of SARS-CoV-2 in Boston Highlights the Impact of Superspreading Events." *Science (New York, N.Y.)* 371 (6529). https://doi.org/10.1126/science.abe3261.

Oude Munnink, B. Bas, E. Münger, D. F. Nieuwenhuijse, R. Kohl, A. van der Linden, C. M. E. Schapendonk, H. van der Jeugd, et al. 2020. "Genomic Monitoring to Understand the Emergence and Spread of Usutu Virus in the Netherlands, 2016–2018." *Scientific Reports* 10 (1): 2798. https://doi.org/10.1038/s41598-020-59692-y.

Paraskevis, D., E. G. Kostaki, G. Magiorkinis, G. Panayiotakopoulos, G. Sourvinos, and S. Tsiodras. 2020. "Full-Genome Evolutionary Analysis of the Novel Corona Virus (2019-NcoV) Rejects the Hypothesis of Emergence as a Result of a Recent Recombination Event." *Infection, Genetics and Evolution: Journal of Molecular Epidemiology and Evolutionary Genetics in Infectious Diseases* 79 (April): 104212. https://doi.org/10.1016/j.meegid.2020.104212.

Payne, Susan. 2017. "Family Coronaviridae." *Viruses*. Elsevier: 149–58. https://doi.org/10.1016/B978-0-12-803109-4.00017-9.

Phan, Tung. 2020a. "Novel Coronavirus: From Discovery to Clinical Diagnostics." *Infection, Genetics and Evolution: Journal of Molecular Epidemiology and Evolutionary Genetics in Infectious Diseases* 79 (April): 104211. https://doi.org/10.1016/j.meegid.2020.104211.

Phan, Tung. 2020b. "Genetic Diversity and Evolution of SARS-CoV-2." *Infection, Genetics and Evolution: Journal of Molecular Epidemiology and Evolutionary Genetics in Infectious Diseases* 81 (July): 104260. https://doi.org/10.1016/j.meegid.2020.104260.

Prado-Vivar, Belen, Monica Becerra-Wong, Juan Jose Guadalupe, Sully Marquez, Bernardo Gutierrez, Patricio Rojas-Silva, Michelle Grunauer, Gabriel Trueba, Veronica Barragan, and Paul Cardenas. 2020. "COVID-19 Re-Infection by a Phylogenetically Distinct SARS-CoV-2 Variant, First Confirmed Event in South America." *SSRN Scholarly Paper ID 3686174*. Rochester, NY: Social Science Research Network. https://papers.ssrn.com/abstract=3686174.

"Preliminary Genomic Characterisation of an Emergent SARS-CoV-2 Lineage in the UK Defined by a Novel Set of Spike Mutations—SARS-CoV-2 Coronavirus/NCoV-2019 Genomic Epidemiology." 2020. *Virological*. December 18, 2020. https://virological.org/t/preliminary-genomic-characterisation-of-an-emergent-sars-cov-2-lineage-in-the-uk-defined-by-a-novel-set-of-spike-mutations/563.

Quick, Joshua, Nicholas J. Loman, Sophie Duraffour, Jared T. Simpson, Ettore Severi, Lauren Cowley, Joseph Akoi Bore, et al. 2016. "Real-Time, Portable Genome Sequencing for Ebola Surveillance." *Nature* 530 (7589): 228–32. https://doi.org/10.1038/nature16996.

"Real-Time PCR: Understanding Ct|Thermo Fisher Scientific—IN." n.d. Accessed March 22, 2021. https://www.thermofisher.com/in/en/home/life-science/pcr/real-time-pcr/real-time-pcr-learning-center/real-time-pcr-basics/real-time-pcr-understanding-ct.html.

Tillett, Richard L., Joel R. Sevinsky, Paul D. Hartley, Heather Kerwin, Natalie Crawford, Andrew Gorzalski, Chris Laverdure, et al. 2021. "Genomic Evidence for Re-infection with SARS-CoV-2: A Case Study." *The Lancet. Infectious Diseases* 21 (1): 52–58. https://doi.org/10.1016/S1473-3099(20)30764-7.

To, Kelvin Kai-Wang, Ivan Fan-Ngai Hung, Jonathan Daniel Ip, Allen Wing-Ho Chu, Wan-Mui Chan, Anthony Raymond Tam, Carol Ho-Yan Fong, et al. 2020. "COVID-19 Re-Infection by a Phylogenetically Distinct SARS-Coronavirus-2 Strain Confirmed by Whole Genome Sequencing." *Clinical Infectious Diseases: An Official Publication of the Infectious Diseases Society of America*, August. https://doi.org/10.1093/cid/ciaa1275.

Van Elslande, Jan, Pieter Vermeersch, Kris Vandervoort, Tony Wawina-Bokalanga, Bert Vanmechelen, Elke Wollants, Lies Laenen, et al. 2020. "Symptomatic SARS-CoV-2 Re-infection by a Phylogenetically Distinct Strain." *Clinical Infectious Diseases: An Official Publication of the Infectious Diseases Society of America*, September. https://doi.org/10.1093/cid/ciaa1330.

Voloch, Carolina M., Ronaldo da Silva Francisco, Luiz G. P. de Almeida, Cynthia C. Cardoso, Otavio J. Brustolini, Alexandra L. Gerber, Ana Paula de C. Guimarães, et al. 2021. "Genomic Characterization of a Novel SARS-CoV-2 Lineage from Rio de Janeiro, Brazil." *Journal of Virology*, March. https://doi.org/10.1128/JVI.00119-21.

Yin, Changchuan. 2020. "Genotyping Coronavirus SARS-CoV-2: Methods and Implications." *Genomics* 112 (5): 3588–96. https://doi.org/10.1016/j.ygeno.2020.04.016.

Zhang, Tao, Qunfu Wu, and Zhigang Zhang. 2020. "Probable Pangolin Origin of SARS-CoV-2 Associated with the COVID-19 Outbreak." *Current Biology: CB* 30 (7): 1346–51.e2. https://doi.org/10.1016/j.cub.2020.03.022.

Zhang, Wenjuan, Brian D. Davis, Stephanie S. Chen, Jorge M. Sincuir Martinez, Jasmine T. Plummer, and Eric Vail. 2021. "Emergence of a Novel SARS-CoV-2 Variant in Southern California." *JAMA*, February. https://doi.org/10.1001/jama.2021.1612.

3

Mass spectrometry techniques for detection of COVID-19 viral and host proteins using naso-oropharyngeal swab and plasma

Harsh Khatri and Kruthi Suvarna

CONTENTS

List of abbreviations

ABC – Ammonium bicarbonate
DDA – Data-dependent acquisition
ESI – Electrospray ionization
FA – Formic acid
HCD – High-energy collision dissociation
IAA – Iodoacetamide
LC – Liquid chromatography
LFQ – Label-free quantification
RT-PCR – Real-time polymerase chain reaction
SDS – Sodium dodecyl sulfate
TCEP – Tris(2-carboxyethyl) phosphine
TMT – Tandem mass tag

DOI: 10.1201/9781003220787-3

3.1 Introduction

The rapid advances in high-throughput proteomics technology have made a significant contribution in understanding the outbreak of several viral infections such as HIV, ZIKV, DENV, MERS-CoV, and SARS-CoV. The mass spectrometry–based proteomics strategies have been very useful in understanding the SARS-CoV-2 virus and host proteins interaction, pathogenesis, and identification of antiviral drug targets (Sperk et al. 2020). The most commonly used strategies for the mass spectrometry–based discovery proteomics are bottom-up and top-down approaches. Compared to the top-down proteomics approach of detecting the intact proteins, bottom-up proteomics has been successfully applied for the precise identification of peptides in complex samples (Leidy 2011). The bottom-up proteomics approach utilizes the digested peptides extracted from the complex biological samples, followed by the liquid chromatographic (LC) elution of peptides which are subjected to electrospray ionization (ESI). The tandem mass spectrometry, also known as MS/MS, allows two-level detection of the peptide ions (precursor ions). During the first level of detection (MS1), the analyzer measures the mass-to-charge ratio of precursor ions and the second level of detection (MS2) involves the measurement of further fragmented peptide ions (product ions). Such complete analysis of the fragmented peptide ions is termed LC-MS/MS (Pitt 2009). Out of several bioanalytical testing measures employed to contain the COVID-19 pandemic, nucleic acid detection techniques such as RT-PCR have been the front-runners for the diagnosis (Hosseini et al. 2020). Although RT-PCR is the gold standard for the detection of viral infection, researchers have also employed a quantitative LC-MS/MS proteomics approach for detection of SAR-CoV-2 viral peptides from complex samples such as blood (serum, plasma, PBMCs), urine, saliva, swab as well as SAR-CoV-2 infected cell pellets (Mahmud and Garrett 2020). The swab samples collected from the primary site of infection have a high viral load and are thus used clinically to detect COVID-19 by RT-PCR. However, considering that swab samples are highly infectious, most of the biochemical tests in the clinics use blood-based samples. Thus, blood samples are considered more suitable for the identification of protein biomarkers for COVID-19 prognosis. Compared to RT-PCR, which can detect specific viral gene targets, the quantitative proteomics-based MS technique has the advantage of identifying nearly all the pathogens from complex clinical samples. Thus, MS can also be multiplexed for the detection of several infections at the same time, making it more suitable for rapid diagnosis. Recently, MS techniques have also been applied for studying the host response, immune evasion, and pathogenesis of SAR-CoV-2. The host proteins altered significantly in response to COVID-19 infection will also aid in the antiviral drug development and the design of therapeutic approaches. This chapter summarizes the recently gathered knowledge, challenges, and opportunities for the mass spectrometry–based biomarker discovery and future potential for clinical application.

3.2 Sample preparation for mass spectrometry analysis

The handling of any infectious virus samples, including SARS-CoV-2, should be conducted in the Biosafety level 3 (BSL-3) laboratory complying with biosafety guidelines. However, most of the laboratories equipped with mass spectrometry instruments might not be having a BLS-3 facility. In such a case, inactivation of samples using either heat treatment or organic solvents should be performed for the safe processing of the samples. Recently, several researchers have identified the viral and host proteins using oro- and nasopharyngeal swabs (Cardozo et al. 2020; Rivera et al. 2020; Akgun et al. 2020). A few MS-based studies have relied on a complex sample preparation protocol, using ethanol or isopropanol organic solvent to extract proteins (Cardozo et al. 2020; Nikolaev et al. 2020). Rivera and his group collected the swab samples of infected patients in the sodium dodecyl sulfate (SDS) buffer and performed in-gel digestion to detect proteins (Rivera et al. 2020). Compared to the swab samples, plasma samples are more suitable for the rapid clinical translation of the MS-detected biomarker. In this light, several studies have focused on the plasma samples to detect and understand host protein response toward SARS-CoV-2 infection (Waage et al. 2017; Shu et al. 2020; Messner et al. 2020). The most abundant proteins in the plasma account for around 99% of proteins by weight and this might be an obstacle for detecting medium or low abundant proteins. The depletion strategies such as immune- or affinity-based depletion of abundant proteins can be used; however, these techniques also have a limitation of increased sample handling and co-depletion might also occur as many of these abundant proteins have carrier functions (Ignjatovic et al. 2020). Joonho Park et al. reported the suitability of undepleted infected plasma samples to identify host proteins (Park et al. 2020). The schematics of workflow for the swab and plasma sample processing for MS analysis are shown in Figure 3.1.

3.2.1 Plasma sample preparation for MS analysis

In this section, we will discuss the detailed workflow of immune–affinity-based plasma depletion. The detailed description of the sample preparation method is shown in Table 3.1. A 2-ml of whole-blood sample is collected in a sterile vacutainer by clinicians. The blood samples are centrifuged at 3,000 rpm for 10 minutes. The plasma samples are then collected carefully from the top of the centrifuged tube. These samples are then heated at 56°C for 30 minutes for viral inactivation. The samples are aliquoted into several vials and stored at −80°C until further use. The depletion of high abundant proteins such as fibrinogen, haptoglobin, albumin, apolipoprotein A-I, and others are carried out using the Pierce™ 12 abundant protein depletion spin column (ThermoFisher Scientific). Around 15 µl of plasma samples were mixed with the resin of the spin column and incubated under a rocking motion for 1 hour. Around 500 µl samples were eluted by centrifugation 1,500 *g* for 2 minutes and concentrated to 100 µl using SpeedVac vacuum. The quantification of depleted plasma samples was carried out by Bradford assay taking bovine

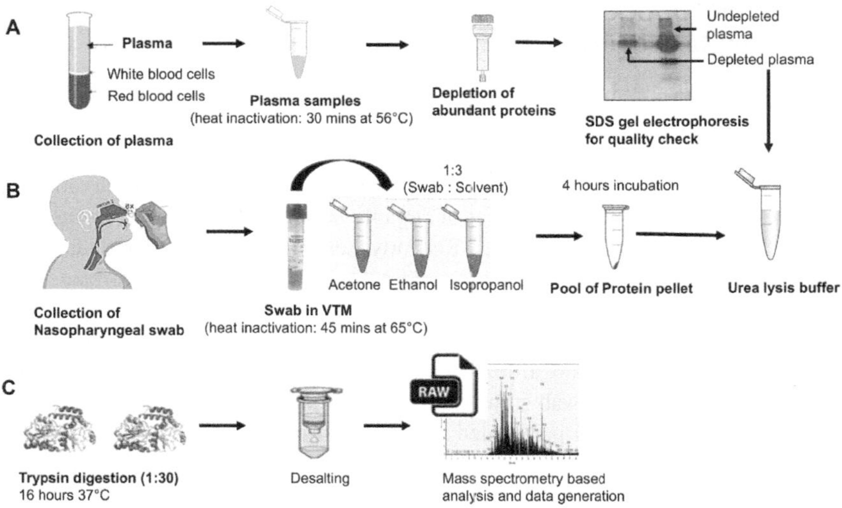

FIGURE 3.1 Schematics for COVID-19-infected swab and plasma sample preparation for mass spectrometry analysis. (a) Schematic for COVID-19 plasma sample preparation. Representation of COVID-19 plasma sample preparation as done by Suvarna et al. (PMID: 33995121) for mass spectrometry analysis. (b) Schematic for COVID-19 swab sample preparation. Representation of COVID-19 swab sample preparation as done by Bankar et al. (PMID: 33558857) for mass spectrometry analysis. (c) Schematic for peptide digestion and analysis using mass spectrometry.

TABLE 3.1

Summary of MS-based Proteomics Analysis Using COVID-19-Infected Swab and Plasma Samples

Mass Spectrometry	Sample Preparation	Labeling	References
Nano-HPLC Dionex UltiMate 3000 system coupled to timsTOF Pro MS	Naso- and oropharyngeal swab samples heat inactivated at 65°C for 30 minutes and treated with isopropanol	Label-free	Nikolaev et al. 2020
Nano LC-MS/MS	Naso- and oropharyngeal swab lyophilized samples	Label-free	Akgun et al. 2020
Nano LC UltiMate 3000 system coupled to Q Exactive HF-X mass spectrometer	Naso- and oropharyngeal swab samples precipitated using ethanol	Label-free	Cardozo et al. 2020
Nano LC UltiMate 3000 system coupled to Q Exactive Plus (Q-Orbitrap) mass spectrometer	Naso-and oropharyngeal swab samples were inactivated by addition of 2% SDS	Label-free	Rivera et al. 2020
Nano-LC coupled to Q Exactive Plus mass spectrometer	Naso-and oropharyngeal swab samples were inactivated by addition of 4% SDS and were kept at 95°C for 15 minutes and protein precipitation was performed in acetone	Label-free	Maras et al. 2020

Nano-HPLC Dionex UltiMate 3000 chromatography system coupled to a Quadrupole-Orbitrap (Q-Orbitrap) mass spectrometer	Nasopharyngeal swab samples were inactivated at 70°C for 1 hour and precipitated using methanol	Label-free	Saadi et al. 2021
Nano-LC system (Eksigent NanoLC-425) coupled to quadrupole TOF (TripleTOF 6600, Sciex, USA) hybrid mass spectrometer	Naso- and oropharyngeal swabs were inactivated by adding lysis buffer (25% guanidinium thiocyanate and 5% SDS) for 20 minutes at RT. Twenty percent (w/v) trichloroacetic acid (TCA) was used for protein precipitation	Label-free	Singh et al. 2020
UltiMate 3000 LC system (Dionex-LC) coupled to Q Exactive HF mass spectrometer (Thermo)	Nasopharyngeal swab was precipitated using trichloroacetic acid and heated at 99°C for 5 minutes in LDS sample buffer	Label-free	Gouveia et al. 2020
Nano-LC 1200 system coupled to Orbitrap Fusion Tribrid Mass Spectrometer (Thermo)	Nasopharyngeal swab was heat inactivated at 65°C for 45 minutes and precipitated using ethanol, acetone, and isopropanol	Label-free	Bankar et al. 2021
Easy-nLC 1200 system coupled to Q Exactive HF-X mass spectrometer	Undepleted plasma mixed with reaction solution (1% SDC, 10 mM TCEP, and 40 mM CAA)	TMT11-plex labeling	Shu et al. 2020
Ultimate 3000 RSLC system (Dionex, USA) coupled to Q Exactive HF-X mass spectrometer	Undepleted plasma	Label-free	Park et al. 2020
High-flow LC-MS systems, an Agilent 1290 Infinity II coupled to a Triple-TOF 6600 mass spectrometer	Undepleted plasma and serum samples	Label-free	Messner et al. 2020
Dionex UltiMate 3000 nano HPLC system coupled to quadrupole-ion trap-Orbitrap hybrid Eclipse® mass spectrometer	Plasma samples were lysed in following lysis buffer (5.4 M guanidinium hydrochloride, 50 mM Tris, 10 mM TCEP, 40 mM chloroacetamide, pH ~ 8)	Label-free	Overmyer et al. 2020
Nanoflow DIONEX UltiMate 3000 RSLCnano System coupled to Q Exactive HF-X hybrid Quadrupole-Orbitrap	Serum samples were inactivated and sterilized at 56°C for 30 minutes	TMTpro16plex label reagents	Shen et al. 2020
Orbitrap Fusion Tribrid Mass Spectrometer (Thermo Fischer Scientific)	Plasma was incubated at 56°C for 30 minutes for viral inactivation. The depleted plasma proteins were reduced using 6 M urea	Label-free	Suvarna et al. 2021

serum albumin (BSA) protein as a reference standard. To 30 µg of a depleted plasma sample was added 6 M of urea. The plasma protein was reduced with 20 mM tris(2-carboxyethyl) phosphine (TCEP) and incubated at 37°C for 1 hour. The alkylation was carried out by 40 mM iodoacetamide (IAA) and incubated for 15 minutes in the dark. Then, the sample was diluted six times with 50 mM ammonium bicarbonate (ABC). The proteins were digested using trypsin at an enzyme/substrate ratio of 1:30 and incubated overnight at 37°C. Next day, the digested sample was vacuum dried to stop trypsin activity and reconstituted in 0.1% formic acid (FA). The digested peptide was desalted using a C-18 column, and the eluted dried peptides were reconstituted in 0.1% (v/v) FA. Finally, the peptide concentration was calculated from its OD value at 205 and 280 nm using the Scopes method (Scopes 1974). Around 1 µg of peptides was injected at a flow rate of 300 nl/minute into liquid chromatography (LC) and the peptides were analyzed on Orbitrap Fusion Tribrid Mass Spectrometer (Thermo Fischer Scientific) with an easy nano LC-1200 system (Suvarna et al. 2021).

3.2.2 Swab sample preparation for MS analysis

Most research works have used different solvents for precipitation of the swab samples, such as acetone, ethanol, methanol, trichloroacetic acid, isopropanol, and SDS. Bankar et al. precipitated swab samples using three solvents—ethanol, isopropanol, and acetone. The pooling of the peptide from these three conditions gave a high abundance of viral peptides compared to the individual solvent precipitation (Bankar et al. 2021). The details of the nasopharyngeal and oropharyngeal swab sample preparation are given in Table 3.1.

3.3 Discovery proteomics for detection of viral and host proteins

Recently, quantitative discovery proteomics has been widely explored to detect the viral and host peptides from the COVID-19-infected samples (Park et al. 2020; Nikolaev et al. 2020; Demichev et al. 2020). Currently, there are two methods of quantification for MS-based proteomics, label-free (LFQ) and label-based quantification techniques (Bantscheff et al. 2007) (Figure 3.2). In the LFQ approach, the peptides are quantified on the basis of peak intensity and spectral counting measurements. The measurement of intensity occurs at the MS1 level and the quantification of the area under the curve generated from the chromatogram (Latosinska et al. 2015). The quantification using label-based approach is based on the use of stable isotopes such as isotope-coded affinity tag, isobaric tag for relative and absolute quantification (iTRAQ), and tandem mass tag (TMT), a chemical labeling technique (Trinh et al. 2013). The label-based methods allow for the labeling as well as quantification of multiple samples. The combining of multiple samples for one run not only saves the instrument time but also allows for reduced analytical variability. A few research studies have used TMT-based chemical labeling techniques to detect host proteins using COVID-19-infected plasma samples (Shen et al. 2020; Shu et al. 2020).

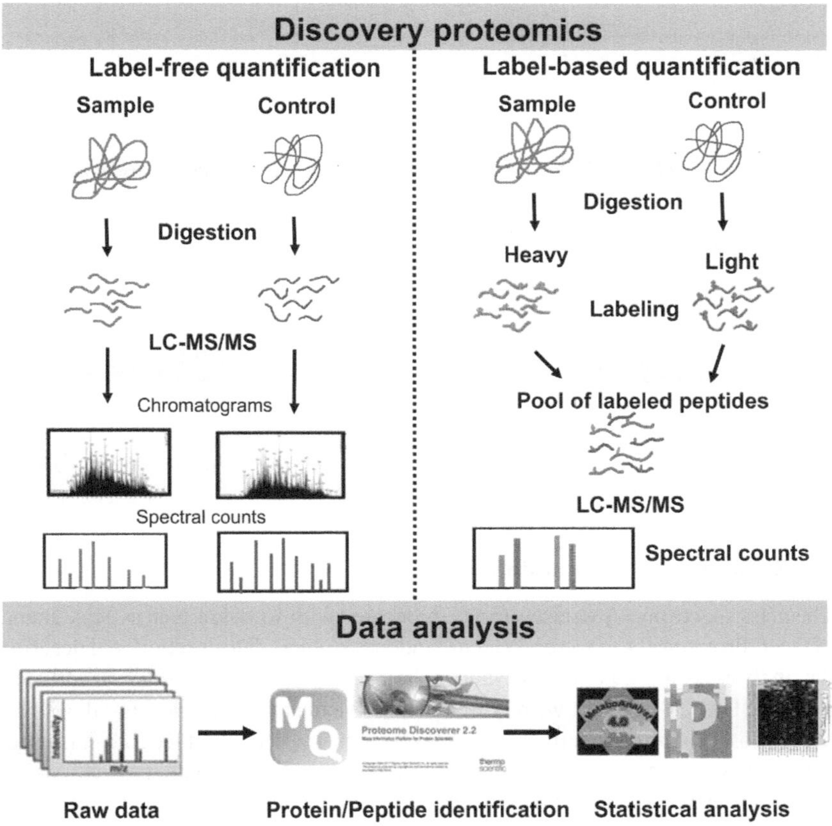

FIGURE 3.2 Workflow for mass spectrometry–based label and label-free quantification.

Source: Representative figure of discovery proteomics using label-based and label-free quantification (LFQ) as described by Deracinois et al. PMID: 28250403.

3.3.1 Label-free quantification

The swab and plasma peptides were analyzed on Orbitrap Fusion Mass Spectrometer with an easy nano-LC 1200 system (Suvarna et al. 2021). The analysis was performed in the DDA (data-dependent acquisition) mode along with a scan range of 375–1,700 *m/z* and a mass resolution of 60,000. A mass tolerance window was set at 10 ppm, and all the MS/MS data were collected using the high-energy collision dissociation method (HCD). An LC gradient of 120 minutes was used for the separation of the peptides. Solvent A consisted of 0.1% FA in water (LC/MS grade) and solvent B comprised of 80% can in 0.1% FA water (LC/MS grade). For data acquisition, Thermo Xcalibur software (version 4.0) was used. The chromatograms generated by mass spectrometry analysis of plasma and swab samples are shown in Figure 3.3.

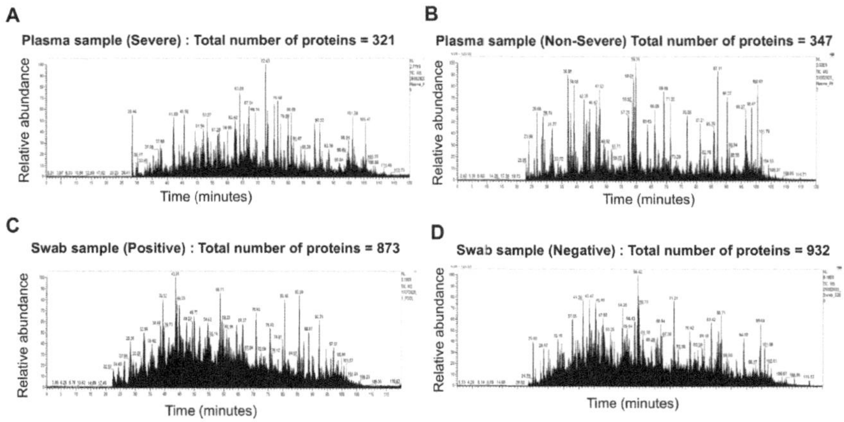

FIGURE 3.3 Representative chromatograms generated by mass spectrometry analysis of plasma and swab samples from COVID-19-infected patients.

3.3.2 Data analysis and database

The mass spectrometry datasets can be processed with software such as MaxQuant, Mascot, Proteome Discoverer, Perseus, and many others. This section will describe the analysis using MaxQuant (Tyanova, Temu, and Cox 2016) (v1.6.6.0) software in detail. For any analysis, we required database and parameter files. For the detection of host proteins, the human Swiss-Prot database (Version: 2020_04) was used, consisting of 20,353 proteins. To detect viral peptides, severe acute respiratory syndrome coronavirus 2 (SARS-CoV-2) Swiss-Prot database consisting of 13 proteins was used. Raw files were processed in MaxQuant using label-free quantification (LFQ) method and setting label-type as "standard" with a multiplicity of 1. The Orbitrap was set to Orbitrap Fusion mode. The parameter file setting was as follows: digestion of protein trypsin was used with a maximum missed cleavage of 2. The fixed modification was set as Carbamido-methylation of cysteine (+57.021464 Da) and variable modification was set as oxidation of methionine (+15.994915 Da). The false discovery rate (FDR) was set to 1% for the protein and peptide levels. Decoy mode was set to "revert", and the type of identified peptides was set to "unique+razor" (Suvarna et al. 2021).

3.4 MS-based detection of host and viral proteins using COVID-19-infected swab samples

Recently, MS-based proteomics techniques have been explored for the detection of both viral and host proteins. The list of host and viral proteins detected in swab samples using MS techniques are shown in Table 3.2. Nikolaev et al. identified around 1,500 proteins in the nasopharynx swabs using an untargeted LC-MS/MS approach, out of which nucleoprotein (P0DTC9) from SARS-CoV-2 was detected with high confidence (Nikolaev et al. 2020). Akgun et al., using nanoLC-MS,

TABLE 3.2
Summary of Viral and Host Proteins Identified Using Swab Samples

Viral Proteins	Host Proteins	Reference
P0DTC9 (NCAP_SARS2)	Over 1,000 proteins detected	Nikolaev et al. 2020
Not studied	Over 207 proteins identified P05164 (MPO), P24158 (PRTN3), P08246 (ELANE), P08311 (CTSG), and P20160 (AZU1)	Akgun et al. 2020
P0DTC2 (SPIKE_WCPV) P0DTC3 (AP3A_WCPV) P0DTC4 (VEMP_WCPV) P0DTC5 (VME1_WCPV) P0DTC7 (NS7A_WCPV) P0DTC8 (NS8_WCPV) P0DTC9 (NCAP_WCPV) P0DTD1; P0DTC1 (R1AB_WCPV; R1A_WCPV)	Not studied	Cardozo et al. 2020
P0DTC9 (NCAP_SARS2)	Over 1,100 proteins identified P32455 (GBP1), D7RIG5 (HLA-DRB1), Q9BX59 (TAPBPL)	Rivera et al. 2020
P0DTC9 (NCAP_SARS2), P59596 (VME1_SARS) P0DTC2 (SPIKE_SARS2) P0DTD1 (R1AB_SARS2) P0DTD2 (ORF9B_SARS2) A0A6H1PRW5 (A0A6H1PRW5_SARS2)	Over 1,256 proteins identified P20591 (MX1), P23381 (WARS1) P07737 (PFN1), Q16630 (CPSF6) P25815 (S100P)	Maras et al. 2020
P0DTC9 (NCAP_SARS2), P0DTC2 (SPIKE_SARS2) P59596 (VME1_SARS)	Not studied	Saadi et al. 2021
P0DTC9 (NCAP_SARS2), P0DTC2 (SPIKE_SARS2) P0DTD1 (R1AB_SARS2)	Not studied	Singh et al. 2020
P0DTC9 (NCAP_SARS2), P0DTC2 (SPIKE_SARS2) P59596 (VME1_SARS)	Not studied	Gouveia et al. 2020
P0DTC9 (NCAP_SARS2), P0DTC2 (SPIKE_SARS2) P0DTD1 (R1AB_SARS2)	P05231 (IL-6), P02741 (CRP), P00338 (LDHA), P07195 (LDHB), P02794 (FTH1), P02792 (FTL), P00505 (GOT2), P17174 (GOT1), P42224 (STAT1)	Bankar et al. 2021

identified 17 statistically significant proteins out of a total of 207 host proteins detected in the swab samples. The significant proteins such as Neutrophil Elastase (ELANE), Azurocidin (AZU1), Myeloperoxidase (MPO), Myeloblastin (PRTN3), Cathepsin G (CTSG) and Transcobalamine-1 (TCN1) were identified to be linked to alteration in the innate immune system, specifically via neutrophil degranulation

(Akgun et al. 2020). Cardozo et al. developed a high-throughput targeted proteomics assay to detect COVID-19 nucleoprotein peptides from nasopharyngeal and oropharyngeal swab samples. Data-dependent acquisition (DDA) analyses revealed around 119 unique peptides belonging to the eight SARS-CoV-2 proteins, out of which nucleoprotein (NCAP_WCPV) accounted for 23.5% of the identified peptides (Cardozo et al. 2020). Rivera et al. detected an average of 1,100 host proteins from the infected swab sample. The most abundant proteins detected in positive swabs were guanylate-binding protein 1, tapasin, and HLA class II histocompatibility antigen DR beta chain. The biological processes altered in infected host cells were SRP-dependent cotranslational protein targeting membrane, viral transcription and translational initiation and nuclear-transcribed mRNA catabolic process, and nonsense-mediated decay (Rivera et al. 2020). Gouveia et al., with a 20-minute MS acquisition window, were able to identify and quantify several virus-specific peptides. They pointed out that the peptides ADETQALPQR (and its variant forms) and GFYAQGSR from the nucleocapsid protein are of most interest for developing rapid, targeted assays (Gouveia et al. 2020). Bankar et al. identified a few of the host proteins such as interleukin-6, ferritin, l-lactate dehydrogenase, C-reactive protein, and aspartate aminotransferase to be upregulated only in COVID-19-positive patients using targeted proteomics assay. The enriched pathways identified in the host were neutrophil degranulation, interleukin-12 (IL-12) signaling pathways, and mRNA translation of proteins (Bankar et al. 2021). Thus, the following studies indicate that MS-detected host proteins have a potential for monitoring the disease progression and the panel of these proteins can be used for clinical translation.

3.5 MS-based detection of host proteins using COVID-19-infected plasma samples

Shu et al. performed proteomics analysis of infected plasma of a cohort of COVID-19 patients, recovered from mild or severe symptoms, and revealed significant alterations of host proteins. Using a machine learning–based approach, their group identified around 11 proteins as biomarkers that distinguished and predicted COVID-19 outcomes. Some of the identified biomarkers were also validated by enzyme-linked immunosorbent assay (ELISA) using a larger cohort. These significantly altered proteins involved in the pathway, such as immune or inflammatory responses, platelet degranulation and coagulation, and metabolism, contributed to the COVID-19 pathogenesis. Park et al. also performed quantitative proteomic analysis using the BoxCar method and revealed that 91 out of 1,222 quantified proteins were differentially expressed depending on the severity of COVID-19. They found around 76 proteins that could be novel prognostic biomarker candidates and also identified the role of neutrophil activation, complement activation, platelet function, and T-cell suppression, as well as pro-inflammatory factors upstream and downstream of interleukin-6, interleukin-1B, and tumor necrosis factor. Messner et al. developed an ultrahigh-throughput serum and plasma proteomics system to facilitate implementation in regulated clinical laboratories. This low-cost workflow handles up to 180 samples per day, enables high-precision quantification, and reduces batch effects for large-scale studies. Using such technology, they identified around 27 potential biomarkers depending on the WHO severity grade of COVID-19. This includes complement factors, the coagulation system, inflammation modulators, and pro-inflammatory factors upstream and downstream of interleukin-6. The host proteins detected in several studies are summarized in Table 3.3.

TABLE 3.3

Summary of Host Proteins Identified Using Plasma/Serum Samples

Host Proteins	Dominant Pathways	Reference
Over 860 proteins detected Upregulated proteins (severe): ORM1, ORM2, S100A9, CRP, AZGP1, CFI, SERPINA3/ACT, and LCP1/LPL Downregulated proteins (severe): FETUB, CETP, and PI16	Inflammation, immune cell migration and platelet degranulation, complement system, coagulation cascades, and energy metabolism	Shu et al. 2020
Over 1,222 proteins detected Upregulated proteins (severe): CRP, SAA1, CFB, Cofilin-1 CFL1, C2, LRG1, APOC1, and TF	Neutrophil activation, complement activation, platelet function, and T-cell suppression	Park et al. 2020
Over 297 proteins detected: A1BG, ACTG1, ALB, APOA1, APOC1, C1R, C1S, C8A, CD14, CFB, CFH, CFI, CRP, FGA, FGB, FGG, GSN, HP, ITIH3, ITIH4, LBP, LGALS3BP, LRG1, SAA1, SAA1; SAA2, SERPINA10, and TF	Complement factors, coagulation system, inflammation modulators, and pro-inflammatory factors upstream and downstream of interleukin 6	Messner et al. 2020
Over 517 proteins detected Upregulated in severe: PRTN3, LCN2, CD24, BPI, CTSG, DEFA1, DEFA4, MMP8, MPO, AGT, FBLN5, NID1, SERPINB1, HRG, VWF, S100A8, S100A9, SAA1, SAA2, and FGG	Neutrophil degranulation, vessel damage, platelet activation and degranulation, blood coagulation, and acute-phase response	Overmyer et al. 2020
Over 894 proteins detected Upregulated proteins: APOA1, APOA2, APOH, APOL1, APOD, and APOM; SAA1, SAA2, SAA4, CRP, SERPINA3, SAP/APCS; ORM1, C6 Downregulated proteins: PPBP, PF4	Dysregulation of macrophage and lipid metabolism, platelet degranulation, complement system pathways, and massive metabolic suppression	Shen et al. 2020
Over 493 proteins detected: SERPINA1, SERPINA3, SERPINF2, CPB2/TAFI, CFH, CFI, CST3, DEFA1, LRG1, LYZC, HBA and HBB, and CA1	Complement and coagulation cascades, immunoglobulins and antimicrobial enzymes, markers of hemolysis and cell lysis	D'Alessandro et al. 2020

(Continued)

TABLE 3.3 (Continued)

Host Proteins	Dominant Pathways	Reference
Around 307 significant protein changes were observed in the transition from healthy donors to COVID-19 mild infections	Sharp difference between mild and moderate cases is observed in peripheral immune cells	Su et al. 2020
Around 1,200 proteins detected: SERPINA4, serum amyloid P-component (APCS), Protein S100-A8 (S100A8), FGG, SERPINA6, and SERPINA3	Regulation of peptidase activity, regulated exocytosis, blood coagulation, complement activation, leukocyte activation involved in immune response, and response to glucocorticoid biological processes	Suvarna et al. 2021
Around 323 proteins detected: Upregulated proteins: DBH, SHGB, TF, ICAM2, THBS1, and C1RL) Downregulated proteins: APCS and ORM1	Complement system activation, infectious disease processes, and natural killer cell-mediated cytotoxicity	Liu et al. 2021
Upregulated proteins (mild versus severe) IREB2, GELS, POLR3D, PON1, and ULBP6 Upregulated proteins (critical) Gal-10, myoglobin, ASGR2, COPE, and VSTM4	Complement activation, blood coagulation, antimicrobial humoral response, acute inflammatory response, and endopeptidase inhibitor activity	Flora et al. 2021
Around 894 proteins identified Upregulated proteins: CK2, SYK, JAK3, TYK2, and IL-12	Hyperactive T-cell and B-cell signaling, compromised innate immune response, and dysregulated inflammation, coagulation, metabolism, RNA splicing, transcription and translation pathways	Kaneko et al. 2021
Longitudinal study Upregulated proteins (indicator of poor prognosis) AGT, B2M, C1R, CFD, SERPINA3, CRP, CST3 Downregulated proteins (indicator of poor prognosis) AHSG, HRG, PLG, and ITIH2	Age-specific molecular response to COVID-19 involving increased inflammation and lipoprotein dysregulation in older patients	Demichev et al. 2020
Longitudinal study Downregulated proteins (early time point): CRP, SAA1, CD14, LBP, and LGALS3BP Upregulated proteins (later time course): APOH, FN1, HRG, KNG1, PLG APOA1, APOC1, APOC2, APOC3, PON1	Enzymatic activity, regulation of plasma protease inhibitors, coagulation factors, and the complement system	Geyer et al. 2021
Longitudinal study (COVID-19 survivors and non-survivors) Upregulated in survivors: AHSG, FETUB, HRG, KNG1, ITIH1, and ITIH2	Not studied	Völlmy et al. 2021

Recently, many researchers have also focused on the longitudinal collection of the blood samples from the COVID-19 patients at different time points. The longitudinal study of the COVID-19-infected patients during different stages of disease progression might give us essential clues on host immune response at an individual level. Demichev et al. studied the time-dependent progression of COVID-19 by collecting plasma samples at 687 sampling points from a cohort of 139 hospitalized patients. They identified around 113 proteins and 55 diagnostic parameters that were dysregulated in the severe patients. These proteins were found to be mediators of inflammation, complement cascade, and include several apolipoproteins (Demichev et al. 2020). Völlmy et al. also compared the plasma proteome of COVID-19 survivors and non-survivors at different time points. They identified key proteins such as AHSG, FETUB, HRG, and KNG1 to be upregulated in the survivors. Such a panel of proteins may aid in the prediction of mortality risk (Völlmy et al. 2021). Further to this work, Geyer et. analyzed 458 longitudinal serum samples from the COVID-19-infected patients. Their work revealed that the proteins such as CRP, SAA1, CD14, LBP, and LGALS3BP were downregulated in the early time course of infection. The regulators of coagulation (APOH, FN1, HRG, KNG1, PLG) and lipid homeostasis (APOA1, APOC3, PON1) were upregulated during the disease progression (Geyer et al. 2021). Such studies provide comprehensive information by identifying key marker proteins for the prediction of disease outcomes as well as provide the knowledge for personalized therapy.

3.6 Conclusions

The strength and advantages of MS techniques can be utilized for the in-depth understanding of the COVID-19 pandemic and pathogenesis. Currently, there are several challenges in the application of MS-based technology in the clinical setting. Most of the MS instruments are high cost to obtain and maintain in the clinics for high-throughput screening of infectious samples. Hence, the development of portable MS can be important for clinical application and the potential for remote screening (Mahmud and Garrett 2020). With improved sensitivity, miniaturized MS can have multiple benefits and applications in COVID-19 as well as other disease outbreaks. Recently, multiplex assays have gained increasing interest for the detection of a panel of proteins that can diagnose and distinguish multiple infections. Researchers have already started exploring the use of such an assay for COVID-19 biomarker discovery. Gisby et al. used Olink proximity extension immunoassays for plasma and serum proteomics measurements. The multiplex immunoassay was used to measure 436 proteins from the longitudinal samples collected from the infected patients (Gisby et al. 2020). Haljasmägi et al. performed a longitudinal study of 92 inflammation-related plasma protein biomarkers using the Proximity Extension Assay technique. This study identified the early inflammation and apoptotic pathways in the COVID-19 severe patients (Haljasmägi et al. 2020). Thus, in addition to the high-throughput MS techniques, the multiplex platform can also be explored for rapid clinical translation and it might be beneficial for studying the multiple infections in an individual.

REFERENCES

Akgun, Emel, Mete Bora Tuzuner, Betul Sahin, Meltem Kilercik, Canan Kulah, Hacer Nur Cakiroglu, Mustafa Serteser, Ibrahim Unsal, and Ahmet Tarik Baykal. 2020. "Proteins Associated with Neutrophil Degranulation Are Upregulated in Nasopharyngeal Swabs from SARS-CoV-2 Patients." *PLoS ONE* 15 (10 October): 1–10. https://doi.org/10.1371/journal.pone.0240012.

Bankar, Renuka, Kruthi Suvarna, Saicharan Ghantasala, Arghya Banerjee, Deeptarup Biswas, Manisha Choudhury, P. Viswanthram, Akanksha Salkar, Ayushi Verma, Avinash Singh, Amrita Mukherjee, Medha Gayathri J. Pai, Jyotirmoy Roy, Alisha Srivastava, Apoorva Bad, and Sanjeeva Srivastava. 2021. "Proteomic Investigation Reveals Dominant Alterations of Neutrophil Degranulation and MRNA Translation Pathways in COVID-19 Patients." *ISCIENCE* 24 (3): 102135. https://doi.org/10.1016/j.isci.2021.102135.

Bantscheff, Marcus, Markus Schirle, Gavain Sweetman, Jens Rick, and Bernhard Kuster. 2007. "Quantitative Mass Spectrometry in Proteomics: A Critical Review." *Analytical and Bioanalytical Chemistry* 389 (4): 1017–31. https://doi.org/10.1007/s00216-007-1486-6.

Cardozo, Karina Helena Morais, Adriana Lebkuchen, Guilherme Gonçalves Okai, Rodrigo Andrade Schuch, Luciana Godoy Viana, Aline Nogueira Olive, Carolina dos Santos Lazari, et al. 2020. "Establishing a Mass Spectrometry-Based System for Rapid Detection of SARS-CoV-2 in Large Clinical Sample Cohorts." *Nature Communications* 11 (1): 1–13. https://doi.org/10.1038/s41467-020-19925-0.

D'Alessandro, Angelo, Tiffany Thomas, Monika Dzieciatkowska, Ryan C. Hill, Richard O. Francis, Krystalyn E. Hudson, James C. Zimring, Eldad A. Hod, Steven L. Spitalnik, and Kirk C. Hansen. 2020. "Serum Proteomics in COVID-19 Patients: Altered Coagulation and Complement Status as a Function of IL-6 Level." *Journal of Proteome Research* 19 (11): 4417–27. https://doi.org/10.1021/acs.jproteome.0c00365.

Demichev, Vadim, Pinkus Tober-lau, Tatiana Nazarenko, Charlotte Thibeault, Harry Whitwell, Oliver Lemke, Annika Röhl, et al. 2020. "A Time-Resolved Proteomic and Diagnostic Map Characterizes COVID-19 Disease Progression and Predicts Outcome." *medRxiv*.

Flora, Daniele C., Aline D. Valle, Heloisa A. B. S. Pereira, Thais F. Garbieri, Nathalia R. Buzalaf, Fernanda N. Reis, Larissa T. Grizzo, et al. 2021. "Quantitative Plasma Proteomics of Survivor and Non-Survivor COVID-19 Patients Admitted to Hospital Unravels Potential Prognostic Biomarkers and Therapeutic Targets." *MedRxiv*. https://doi.org/10.1101/2020.12.26.20248855.

Geyer, Philipp E., Florian M. Arend, Sophia Doll, Marie-luise Louiset, Sebastian Virreira Winter, Johannes B. Müller-Reif, Furkan M. Torun, et al. 2021. "High-Resolution Longitudinal Serum Proteome Trajectories in COVID-19 Reveal Patients-Specific Seroconversion Graphical Abstract High-Resolution Longitudinal Serum Proteome Trajectories in COVID-19 Reveal Patients-Specific Seroconversion." *MedRxiv*. https://doi.org/10.1101/2021.02.22.21252236.

Gisby, Jack, Candice L. Clarke, Nicholas Medjeral-Thomas, Talat H. Malik, Artemis Papadaki, Paige M. Mortimer, Norzawani B. Buang, et al. 2020. "Longitudinal Proteomic Profiling of High-Risk Patients with COVID-19 Reveals Markers of Severity and Predictors of Fatal Disease." *MedRxiv*. https://doi.org/10.1101/2020.11.05.20223289.

Gouveia, Duarte, Guylaine Miotello, Fabrice Gallais, Jean Charles Gaillard, Stéphanie Debroas, Laurent Bellanger, Jean Philippe Lavigne, et al. 2020. "Proteotyping SARS-CoV-2 Virus from Nasopharyngeal Swabs: A Proof-of-Concept Focused on a 3 Min Mass Spectrometry Window." *Journal of Proteome Research* 19 (11): 4407–16. https://doi.org/10.1021/acs.jproteome.0c00535.

Haljasmägi, Liis, Ahto Salumets, Anna Pauliina Rumm, Meeri Jürgenson, Ekaterina Krassohhina, Anu Remm, Hanna Sein, et al. 2020. "Longitudinal Proteomic Profiling Reveals Increased Early Inflammation and Sustained Apoptosis Proteins in Severe COVID-19." *Scientific Reports* 10 (1): 1–12. https://doi.org/10.1038/s41598-020-77525-w.

Hosseini, Amin, Richa Pandey, Enas Osman, Amanda Victorious, Feng Li, Tohid Didar, and Leyla Soleymani. 2020. "Roadmap to the Bioanalytical Testing of COVID-19: From Sample Collection to Disease Surveillance." *ACS Sensors* 5 (11): 3328–45. https://doi.org/10.1021/acssensors.0c01377.

Ignjatovic, Vera, Philipp E. Geyer, Krishnan K. Palaniappan, Jessica E. Chaaban, Gilbert S. Omenn, Mark S. Baker, Eric W. Deutsch, and Jochen M. Schwenk. 2020. "Mass Spectrometry-Based Plasma Proteomics: Considerations from Sample Collection to Achieving Translational Data." *Journal of Proteome Research* 18 (12). HHS Public Access: 4085–97. https://doi.org/10.1021/acs.jproteome.9b00503.Mass.

Kaneko, Tomonori, Sally Esmail, Courtney Voss, Claudio M. Martin, and Marat Slessarev. 2021. "System-Wide Hematopoietic and Immune Signaling Aberrations in COVID-19 Revealed by Deep Proteome and Phosphoproteome Analysis." *medRxiv*.

Latosinska, Agnieszka, Konstantinos Vougas, Manousos Makridakis, Julie Klein, William Mullen, Mahmoud Abbas, Konstantinos Stravodimos, et al. 2015. "Comparative Analysis of Label-Free and 8-Plex ITRAQ Approach for Quantitative Tissue Proteomic Analysis." *PLoS ONE* 10 (9): 1–25. https://doi.org/10.1371/journal.pone.0137048.

Leidy. 2011. "基因的改变NIH Public Access." *Bone* 23 (1): 1–7. https://doi.org/10.1021/cr3003533.Protein.

Liu, Xiaoling, Yinghao Cao, Hongmei Fu, Jie Wei, Jianhong Chen, Jun Hu, and Bende Liu. 2021. "Proteomics Analysis of Serum from COVID-19 Patients." *ACS Omega* 6 (11): 7951–58. https://doi.org/10.1021/acsomega.1c00616.

Mahmud, Iqbal, and Timothy J. Garrett. 2020. "Mass Spectrometry Techniques in Emerging Pathogens Studies: COVID-19 Perspectives." *Journal of the American Society for Mass Spectrometry* 31 (10): 2013–24. https://doi.org/10.1021/jasms.0c00238.

Maras, Jaswinder Singh, Shvetank Sharma, Adil Rafiq Bhat, Reshu Aggrawal, Ekta Gupta, and Shiv K. Sarin. 2020. "Multi-Omics Integration Analysis of Respiratory Specimen Characterizes Baseline Molecular Determinants Associated with COVID-19 Diagnosis." *MedRxiv*. https://doi.org/10.1101/2020.07.06.20147082.

Messner, Christoph B., Vadim Demichev, Daniel Wendisch, Laura Michalick, Matthew White, Anja Freiwald, Kathrin Textoris-Taube, et al. 2020. "Ultra-High-Throughput Clinical Proteomics Reveals Classifiers of COVID-19 Infection." *Cell Systems* 11 (1): 11–24.e4. https://doi.org/10.1016/j.cels.2020.05.012.

Nikolaev, Evgeny N., Maria I. Indeykina, Alexander G. Brzhozovskiy, Anna E. Bugrova, Alexey S. Kononikhin, Natalia L. Starodubtseva, Evgeny V. Petrotchenko, Grigoriy I. Kovalev, Christoph H. Borchers, and Gennady T. Sukhikh. 2020.

"Mass-Spectrometric Detection of SARS-CoV-2 Virus in Scrapings of the Epithelium of the Nasopharynx of Infected Patients via Nucleocapsid N Protein." *Journal of Proteome Research* 19 (11): 4393–97. https://doi.org/10.1021/acs.jproteome.0c00412.

Overmyer, Katherine A., Evgenia Shishkova, Ian J. Miller, Joseph Balnis, Matthew N. Bernstein, Jesse G. Meyer, Qiuwen Quan, et al. 2020. "Large-Scale Multi-Omic Analysis of COVID-19 Severity." *medRxiv*.

Park, Joonho, Hyeyoon Kim, So Yeon Kim, Yeonjae Kim, Jee Soo Lee, Kisoon Dan, Moon Woo Seong, and Dohyun Han. 2020. "In-Depth Blood Proteome Profiling Analysis Revealed Distinct Functional Characteristics of Plasma Proteins between Severe and Non-Severe COVID-19 Patients." *Scientific Reports* 10 (1): 1–10. https://doi.org/10.1038/s41598-020-80120-8.

Park, Joonho, Hyeyoon Kim, So Yeon Kim, Yeonjae Kim, Jee Soo Lee, Moon Woo Seong, and Dohyun Han. 2020. "In-Depth Blood Proteome Profiling Analysis Revealed Distinct Functional Characteristics of Plasma Proteins between Severe and Non-Severe COVID-19 Patients." *BioRxiv*. https://doi.org/10.1101/2020.08.18.255315.

Pitt, James J. 2009. "Principles and Applications of Liquid Chromatography-mass Spectrometry in Clinical Biochemistry." *Clinical Biochemist Reviews* 30 (February): 19–34.

Rivera, Bernardina, Alejandro Leyva, María Magdalena Portela, Gonzalo Moratorio, Pilar Moreno, Rosario Durán, and Analía Lima. 2020. "Quantitative Proteomic Dataset from Oro- and Naso-Pharyngeal Swabs Used for COVID-19 Diagnosis: Detection of Viral Proteins and Host's Biological Processes Altered by the Infection." *Data in Brief* 32: 106121. https://doi.org/10.1016/j.dib.2020.106121.

Saadi, Justyna, Saoussen Oueslati, Laurent Bellanger, Fabrice Gallais, Laurent Dortet, Anne-Marie Roque-Afonso, Christophe Junot, Thierry Naas, François Fenaille, and François Becher. 2021. "Quantitative Assessment of SARS-CoV-2 Virus in Nasopharyngeal Swabs Stored in Transport Medium by a Straightforward LC-MS/MS Assay Targeting Nucleocapsid, Membrane, and Spike Proteins." *Journal of Proteome Research* 20 (2): 1434–43. https://doi.org/10.1021/acs.jproteome.0c00887.

Scopes, R. K. 1974. "Measurement of Protein by Spectrophotometry at 205 Nm." *Analytical Biochemistry* 59 (1): 277–82. https://doi.org/10.1016/0003-2697(74)90034-7.

Shen, Bo, Xiao Yi, Yaoting Sun, Xiaojie Bi, Juping Du, Chao Zhang, Sheng Quan, et al. 2020. "Proteomic and Metabolomic Characterization of COVID-19 Patient Sera." *Cell* 182 (1): 59–72.e15. https://doi.org/10.1016/j.cell.2020.05.032.

Shu, Ting, Wanshan Ning, Di Wu, Jiqian Xu, Qiangqiang Han, Muhan Huang, Xiaojing Zou, et al. 2020. "Plasma Proteomics Identify Biomarkers and Pathogenesis of COVID-19." *Immunity* 53 (5): 1108–22.e5. https://doi.org/10.1016/j.immuni.2020.10.008.

Singh, Praveen, Rahul Chakraborty, Robin Marwal, V. S. Radhakrishan, Akash Kumar Bhaskar, Himanshu Vashisht, Mahesh S. Dhar, et al. 2020. "A Rapid and Sensitive Method to Detect SARS-CoV-2 Virus Using Targeted-Mass Spectrometry." *Journal of Proteins and Proteomics* 11 (3): 159–65. https://doi.org/10.1007/s42485-020-00044-9.

Sperk, Maike, Robert Van Domselaar, Jimmy Esneider Rodriguez, Flora Mikaeloff, Beatriz Sá Vinhas, Elisa Saccon, Anders Sönnerborg, et al. 2020. "Utility of Proteomics in Emerging and Re-Emerging Infectious Diseases Caused by RNA Viruses." *Journal of Proteome Research* 19 (11): 4259–74. https://doi.org/10.1021/acs.jproteome.0c00380.

Su, Yapeng, Daniel Chen, Dan Yuan, Christopher Lausted, Jongchan Choi, Chengzhen L. Dai, Valentin Voillet, et al. 2020. "Multi-Omics Resolves a Sharp Disease-State Shift between Mild and Moderate COVID-19." *Cell* 183 (6): 1479–95.e20. https://doi.org/10.1016/j.cell.2020.10.037.

Suvarna, Kruthi, Deeptarup Biswas, Medha Gayathri J. Pai, Arup Acharjee, Renuka Bankar, Viswanthram Palanivel, Akanksha Salkar, Viswanthram Palanivel, Akanksha Salkar, Ayushi Verma, Amrita Mukherjee, Manisha Choudhury, Saicharan Ghantasala, Susmita Ghosh, Avinash Singh, Arghya Banerjee, Apoorva Badaya, Surbhi Bihani, Gaurish Loya, Krishi Mantri, Ananya Burli, Jyotirmoy Roy, Alisha Srivastava, Sachee Agrawal, Om Shrivastav, Jayanthi Shastri, and Sanjeeva Srivastava. 2021. "Proteomics and Machine Learning Approaches Reveal a Set of Prognostic Markers for COVID-19 Severity With Drug Repurposing Potential." *Frontiers in Physiology* 12 (2021): 432.

Trinh, Hung V., Jonas Grossmann, Peter Gehrig, Bernd Roschitzki, Ralph Schlapbach, Urs F. Greber, and Silvio Hemmi. 2013. "ITRAQ-Based and Label-Free Proteomics Approaches for Studies of Human Adenovirus Infections." *International Journal of Proteomics* 2013 (Cid): 1–16. https://doi.org/10.1155/2013/581862.

Tyanova, Stefka, Tikira Temu, and Juergen Cox. 2016. "The MaxQuant Computational Platform for Mass Spectrometry-Based Shotgun Proteomics." *Nature Protocols* 11 (12): 2301–19. https://doi.org/10.1038/nprot.2016.136.

Völlmy, Franziska, Henk Van Den Toorn, Riccardo Zenezini Chiozzi, Ottavio Zucchetti, Alberto Papi, Carlo Alberto Volta, Luisa Marracino, et al. 2021. "Is There a Serum Proteome Signature to Predict Mortality in Severe COVID-19 Patients?" *medRxiv*. https://doi.org/10.1101/2021.03.13.21253510.

Waage et al. 2017. "乳鼠心肌提取 HHS Public Access." *Physiology & Behavior* 176 (1): 139–48. https://doi.org/10.1586/erm.13.24.Application.

4

Targeted proteomic approaches in the context of COVID-19 pandemic

Mehar Un Nissa, Alisha Srivastava, and Medha Gayathri J. Pai

CONTENTS

List of abbreviations

AGT – Angiotensinogen
APOB – Apolipoprotein B
BSA – Bovine serum albumin
COVID-19 – Coronavirus disease 2019
DDA – Data-dependent acquisition
DIA – Data-independent acquisition
ESI – Electrospray ionization
FAIMS – High-field asymmetric ion mobility spectrometry
FDR – False discovery rate
FGG – Fibrinogen gamma chain
HER2 – Human epidermal growth factor receptor 2

DOI: 10.1201/9781003220787-4

LC-MS/MS – Liquid chromatography with tandem mass spectrometry
m/z – Mass to charge ratio
MS – Mass spectrometry
NCAP – Nucleoprotein
PRM – Parallel reaction monitoring
PTMs – Post-translational modifications
Q1 – First quadrupole
Q3 – Third quadrupole
QC – Quality control
RDT – Rapid diagnostic test
RT-PCR – Reverse transcription polymerase chain reaction
RT-PCR – Reverse transcription polymerase chain reaction
SARS-CoV-2 – Severe acute respiratory syndrome coronavirus 2
SEPRING1 – Serpin family G member 1
SERPINA3 – Serpin family A member 3
SRM – Selected reaction monitoring
SWATH – Sequential window acquisition of all theoretical mass spectra
TFC-MS – Turbulent flow chromatography mass spectrometry
ToF – Time of flight
TP – Targeted proteomics
TQ – Triple quadrupole
VC – Virus plus cell

4.1 Introduction

4.1.1 The pandemic

With the release of the complete SARS-CoV-2 viral genome sequence, highly sensitive RT-PCR assays were established and brought into mainstream diagnostics (Zhu et al. 2020; Ren et al. 2020). Although this technique is the gold standard to confirm SARS-CoV-2 infection, the high demand and expensive reagents limit its application for large-scale screening. More sophisticated and faster RT-PCR assays were also introduced and used (van Kasteren et al. 2020; Dramé et al. 2020). Various antibody-based rapid diagnostic tests (RDTs) and immunoassays were also developed (van Elslande et al. 2020; Vandenberg et al. 2020). Although these tests were not as accurate as RT-PCR tests, they offered plausible alternatives for large-scale screening (Chaimayo et al. 2020). While the RDTs and immunoassays are simple, rapid, and reasonably sensitive, they simultaneously face many technical challenges. First, these assays are highly dependent on the purity and activity of the antibody being used. Immunoassays are prone to interference and cross-reactivity, leading to erroneous results. In contrast, the targeted MS-based proteomics assays offer superior alternatives (Cardozo et al. 2020). For the sake of simplicity, in this chapter, the MS-based targeted proteomics assays have been generally termed as targeted proteomics (TP) unless mentioned otherwise.

TP approach, in the field of diagnostics, is not unprecedented. It has been in use for the routine monitoring of well-known clinical biomarkers. One example is thyroglobulin as a tumor biomarker in serum samples (Hoofnagle et al. 2008). There are various other examples in which selected reaction monitoring (SRM) methods for detecting biomarkers specific to prostate cancer, lung cancer, ovarian cancer, breast cancer, etc., were developed (Huillet et al. 2012; Henderson et al. 2017; Park et al. 2020).

4.1.2 Comparing targeted proteomics and discovery-based proteomics

Proteomics approaches can be broadly classified into two types: discovery-based proteomics and TPs. Shotgun proteomics can be considered analogous to count the total number of balls in a ball pit, ordering and categorizing them with respect to their characteristics to get a bigger picture. Meanwhile, one studies and compares a few selected balls picked from the ball pit in targeted analysis. Both start with essentially similar experimental strategies. The biological samples are homogenized and proteins are extracted. These are further separated using chromatography and subsequently analyzed on mass spectrometers. The primary difference between both techniques is how the ions are detected and the type of data being acquired. Discovery-based proteomics involves scouring through the entire proteome or the part of it that is feasible by the instrument. Using discovery approaches is good to explore an unknown or a relatively new pathway/disease/organism. On the other hand, targeted analyses are suitable for hypothesis-driven work with more straightforward data analysis steps. The interpretation of targeted data is fairly easy as it does not require complex bioinformatics.

In terms of sensitivity and reproducibility, an adequately optimized TP technique like SRM or PRM (details in further sections) will give better results. Here, we are instructing the instrument to look for specific proteins. The focus, therefore, will be on these proteins that we are interested in studying, resulting in increased sensitivity and reproducibility of the technique. This forges a way to detect low abundant proteins (Marx 2013). This is not the case in the data-dependent acquisition (DDA) based discovery approach, where there are lesser chances of reproducibility between labs, experiments, and runs (Bell et al. 2009; Tabb et al. 2010; Bruderer et al. 2015).

4.1.3 The Triple Quadrupole mass spectrometer

It can be unarguably said that the field of mass spectrometry–based targeted proteomics has been possible because of the pioneering work on the workhorse Triple Quadrupole. As the name suggests, a Triple Quadrupole (TQ) mass spectrometer has three quadrupoles; the first and third quadrupoles act as "mass filters" while the second one acts as a collision cell. We refer to the quadrupoles as "mass filters" as they can allow ions having specific *m*/*z* values to pass through, hence in a sense "filtering the ions".

The instrument was developed in the late 1970s and has witnessed seminal improvements over the years. The small molecule industry has heavily utilized the instrument. They worked alongside the instrument manufacturers to increase the instrument's efficiency and reproducibility. Before it was employed in proteomics-based approaches,

TQ was primarily used to identify and quantify drug standards and drug metabolites in blood samples. Noticing the immense potential of the instrument, researchers started exploring its applications in the field of proteomics. One of the first groups working on TPs developed a very selective assay to detect peptides from biological tissue (Desiderio and Kai 1983). With simultaneous improvements in the field of mass spectrometry, especially the development of ESI and liquid chromatography, mass spectrometry became more reliable for targeted peptide analysis (Fenn et al. 1989).

4.1.4 The gold standard

The gold standard in targeted techniques is selected reaction monitoring (SRM) or multiple reaction monitoring (MRM). Originally introduced in 2011, the technique utilizes a TQ to monitor and quantify the targeted peptides from a protein mixture (Zhang et al. 2011). The target peptides are monitored using a list of transitions. Transitions are a set of peptides (precursor) ion *m/z* and fragment (product) ion *m/z*. The concept might seem similar to any antibody-based immunoassay, where information related to the target protein is a prerequisite, and many studies have pointed out that antibody-based assays are more sensitive when monitoring the levels of a few proteins (Marx 2013). However, the accuracy of an immunoassay often depends on the purity and specificity of the antibody being used, which could otherwise yield erroneous results. Furthermore, antibody-based assays are not easy to scale up. Many factors like the availability of antibody for the target selected and whenever antibody is present, antibody quality, sensitivity, and cross-reactivity need to be considered. On the other hand, the SRM experiments can efficiently monitor tens or hundreds of proteins in low sample volumes (Reiter et al. 2011; Kopylov et al. 2013; de Graaf et al. 2011). Once an optimized method is made, the process is swift and efficient and can be easily applied for high-throughput analyses. The workflow can easily be followed by researchers, quickly automated, multiplexed, and readily translated to clinical settings. Various studies have reported an extensive assessment to check for the robustness and reproducibility leading to better transferability of this assay (Addona et al. 2009; Sherman et al. 2009).

4.1.5 Other targeted techniques beyond targeted proteomics

With the advent of newer technologies, a few other targeted MS techniques have also come into existence. Here, we discuss parallel reaction monitoring (PRM) and sequential window acquisition of all theoretical mass spectra (SWATH).

PRM or sometimes called high-resolution MRM uses the same concept as MRM, but it monitors the target ions at higher resolution mass analyzers (for example Orbitrap). Here all possible fragments of the target peptides are monitored by the instrument. SWATH is a data-independent acquisition (DIA) technique. Here, the target peptides are fragmented and the resulting ions are monitored in a systematic manner across the mass range. This technique, hence, supports the quantitative analyses of peptides (which cover thousands of proteins) with very high accuracy. Figure 4.1 represents this primary difference in the three TP techniques. Refer to Box 4.1 for a comparative explanation of the three MS techniques.

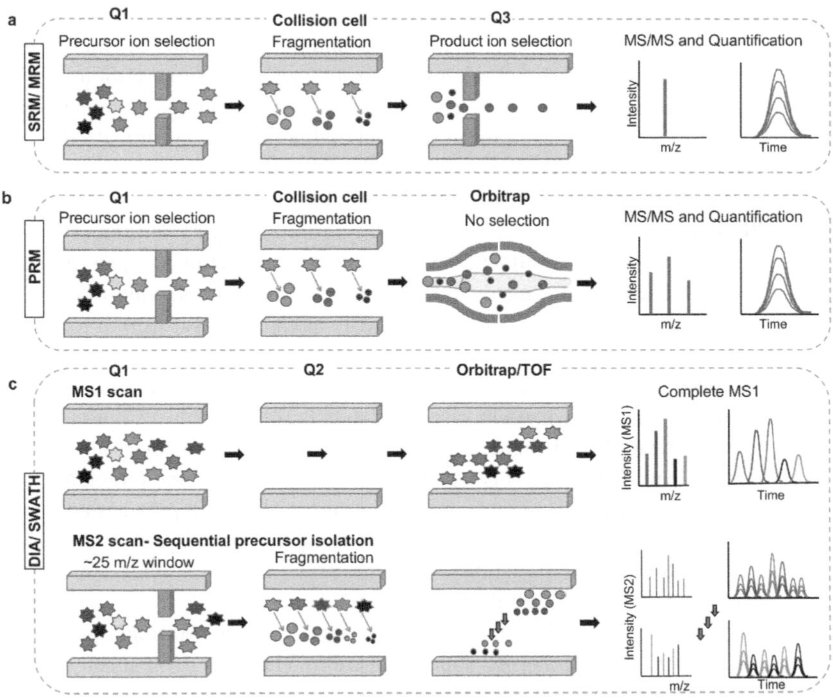

FIGURE 4.1 Representation of principle of targeted proteomics approaches used for detection of target peptides. (a) Schematic for SRM/MRM. Peptides after separation in liquid chromatography are subjected to two steps of mass selection. Targeted precursor ions after getting selected in the first quadrupole (Q1) get fragmented in the collision cell followed by the selection of product ions in the third quadrupole (Q3). (b) Schematic for PRM. Peptides are subjected to only one step of selection where target precursor ion is selected in Q1 and gets fragmented in the collision cell and all resulting fragment ions are parallelly monitored. (c) Schematic for SWATH-MS. First mass analyzer in SWATH-MS is a quadrupole and the second can be time of flight (ToF) or Orbitrap. In the beginning, all peptides eluting at a given time are detected in MS1 scan which is followed by a sequence of MS2 scans depending upon the size of the isolation window. Here five different precursor ions are shown which are monitored in MS1 scan. For MS2, all the precursor ions falling within the window of ~25 *m/z* are sequentially detected. Continuous MS1 and MS2 scans generate data corresponding to all precursors and product ions.

BOX 4.1 SRM/MRM, PRM, AND SWATH

Let us dive deeper into the workings of these three types of targeted techniques by comparing their various pros and cons.

SRM/MRM: An MRM experiment is known to be highly reproducible and precise. When coupled with heavy labeled peptides as internal standards, the technique is highly specific and can be used for absolute quantification. This gold standard in TPs has its demerits. To start with, the prerequisites for the setup are quite tedious. Precursors and fragments that can be detected without interference from one another need to be selected and hence, the optimization takes some time. This also leads to the limited number of proteins or peptides that can be analyzed in one MRM assay.

PRM: Like MRM, PRM also shows high reproducibility and precision and when coupled with labeled peptides spiked into samples, it can be used for absolute quantification easily. It requires quite some effort for the setup and the resulting data analysis is also similar to, if not more complex than MRM. One major benefit here is that for a given precursor, all the possible fragments can be monitored in one go. In cases when our discovery data is from a high-resolution instrument, PRM is beneficial to use. It utilizes the same mass analyzer leading to a better correlation between discovery and targeted data. Also, PRM is slightly more convenient for developing assays for quantifying proteins and peptides than MRM. It is also more suited for the quantification of multiple proteins in a complex sample with attomole-level detection.

SWATH: The major benefit of this DIA method is the minimal optimization or setup before data acquisition. The only prerequisites are the *m/z* values in windows for MS1 and MS2. It has better data coverage than both SRM and PRM. In terms of reproducibility and sensitivity, this wide range of peptides being covered can lead to complications because of interference. Also, the analysis of the acquired data is the most complex among the three, because of the complex nature of the data as well as the analysis pipelines to follow.

4.2 How is a targeted proteomics experiment performed?

One of the most important aspects of carrying out a targeted proteomic experiment is having prior knowledge of the target proteins, their peptides, and post-translational modifications (PTMs) if any. These experiments need to be well thought out and planned for their perfect execution. This includes designing the experiment (which involves deciding on the transitions to be monitored and method preparation), initial data acquisition (preferably using pooled samples), data analysis for optimization of the method and refinement of the transition list based on the acquired data, and optimization of cycle time and dwell time followed by the final data acquisition and analysis using the individual samples. One of the most widely used software for conducting a TP experiment in the proteomics community is Skyline (MacLean et al. 2010; Pino et al. 2020). It can be used to design any targeted experiment, and then to analyze and visualize the results. It can also be used for conducting Metabolomics and Data Independent Acquisition (DIA) experiments.

4.2.1 Designing the experiment

The first step in designing the experiment is creating "Transition Lists". A transition list contains the list of the peptides of the target proteins, and the transitions (precursor and product ions in MRM and only precursor ions in PRM). This transition list needs to be refined keeping in mind the following points.

- Selecting target proteins
- Selecting the best representative peptides of protein. The target peptides must include the following: (i) *Proteotypic peptides*—the best flying peptides per protein; (ii) *Quantotypic peptides*—unique peptides per protein that represent the level of protein (Worboys et al. 2014).
- Selecting optimal transitions per peptide

4.2.2 Data acquisition

For this part of the experiment, the samples are prepared, followed by MS-MS analysis. One can use the appropriate method of protein extraction for their respective sample in question. The extracted protein can be further digested and desalted for the LC-MS/MS runs.

4.2.3 Data analysis

Once data acquisition is completed, the experiment results in the generation of .raw files. These raw files can now be imported to Skyline and analyzed. The process of identifying the right peak of the target peptide is called Peak Annotation and is the most crucial part of data analysis of an SRM/PRM experiment. Several parameters should be considered while annotating peaks:

A. *Co-elution of transitions:* All the transitions of a peptide must co-elute.
B. *Intensity of the peak:* The most intense peaks must be chosen as they are more probable to be corresponding to the target peptide. Also, the peak should be well distinguished from noise signals, if any.
C. *Shape of the peak:* The peaks should be forming a proper Gaussian shape. It is usually helpful to look for the characteristic peak shape.
D. *Retention time prediction:* Skyline can be used to predict the time point at which the target peptide would most likely elute. This predicted retention time can hence be used to identify the peak.
E. *Comparing with spectral library:* If you have built a spectral library, then that can also be very instrumental in identifying the best peak (and hence

the target peptide). A dot product depicts the match of your target peptide spectra with the reference peptide in the library spectra. The greater the dot product, the greater the resemblance between the target peptide and the reference peptide.

F. *Using synthetic peptides:* Synthetic peptides labeled with a heavy isotope can be used to confirm the peak identified is of the right peak.

4.2.4 Ensuring good-quality data

During a targeted experiment, it is essential to ensure that the generated data is unbiased and reproducible. Quality control steps are important, and some standard samples should be run at regular intervals of the experiment. This could be done by monitoring a set of optimized peptides from a pure protein or a heavy labeled synthetic peptide spiked into the samples. The selected peptides should be tested to give a consistent response for the experimental setup. Figure 4.2a represents one peptide of bovine serum albumin (BSA) being used for quality check across five days. Similarly, Figure 4.2b shows the use of a heavy labeled synthetic peptide as spiked in standard to check the quality of response across various samples.

4.2.5 MRM assay for routine clinical analysis

Targeted proteomic assay that can be used for clinical applications involves several other checkpoints after method development. These checkpoints define whether the targeted experiment can actually be used as a routine clinical assay. First, it is of utmost importance that the developed method is highly selective. The matrix of the samples to be analyzed should not interfere or give false positives. Next, to determine in which range quantification can be done, synthetic peptides of the target peptides are run at different concentrations and calibration curves plotted. The limit of detection and limit of quantification are determined. Running several replicates of the samples with known concentration of analyte and synthetic peptide spiked-in also helps in determining the accuracy and precision of the method. The last checkpoint is the stability of these chosen peptides over longer duration of time in the sample of interest. The samples with known analyte are stored and monitored at different time points to check for the effect on quantification values. Thus, optimized and scrutinized targeted method—usually an MRM assay—can be tested on clinical samples.

Andy Hoofnagle and his team developed such a method for quantification of thyroglobulin in serum of cancer patients as early as 2008 (Hoofnagle et al. 2008). This assay detects and quantifies three peptides of the protein thyroglobulin and determines whether the patient needs further therapy. Recently, another such assay has been developed for breast cancer tissue samples. This assay is a more sensitive method of detecting whether a patient is HER2 positive (Kennedy et al. 2021).

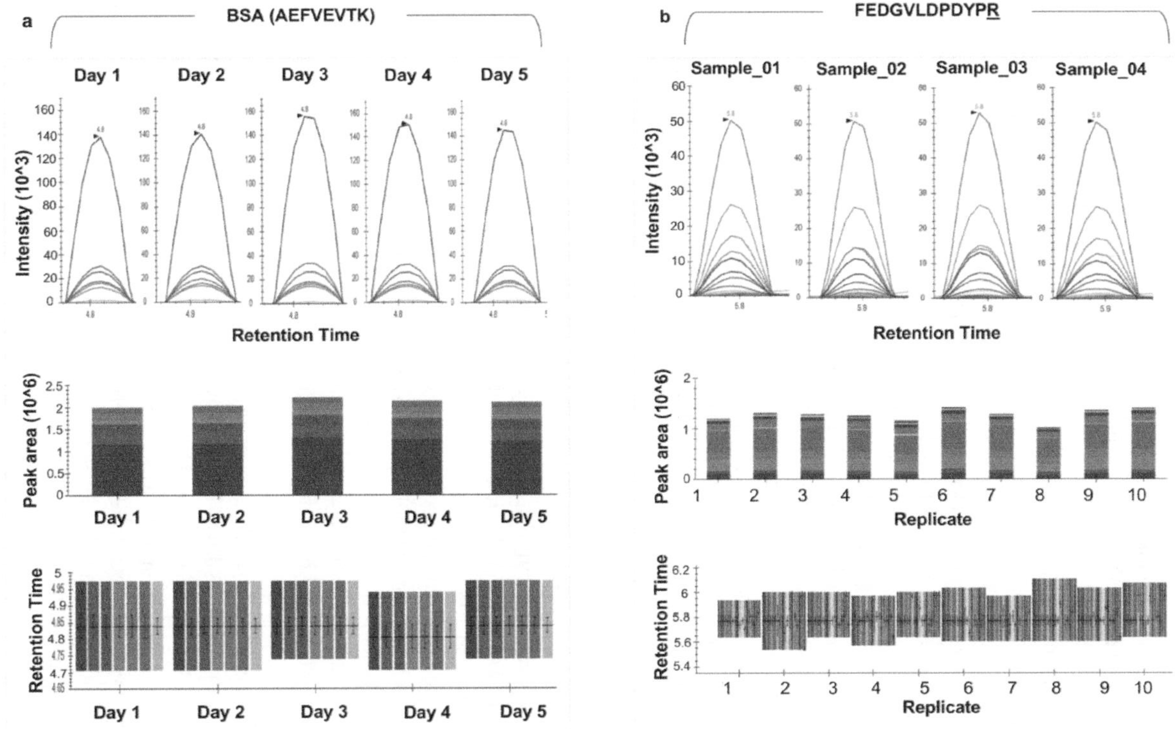

FIGURE 4.2 Quality control checkpoints during an MRM experiment. (a) BSA peptide as instrument QC. Representative peptide of BSA monitored on five consecutive days showing uniformity in peak shape, peak areas, and retention times. (b) Heavy labeled synthetic peptide as an internal standard. Equal amounts of a heavy labeled synthetic peptide are often spiked into the individual samples to monitor the response of the instrument. A few samples of a batch with equal amounts of spiked peptide showing uniform MRM peak shapes, peak areas, and retention times.

4.3 Current targeted proteomic studies for SARS-CoV-2 and authors' perspective

Severe acute respiratory syndrome coronavirus 2 (SARS-CoV-2) infection has spread very rapidly world over, thus exerting huge pressure on the medical system (Grossegesse et al. 2020). This pandemic had brought the entire world to a halt, infecting around 124 million people worldwide. It forced the scientific community to join hands and work together toward developing better prognostics, diagnostics, and treatments for the disease. The proteomics community has contributed immensely toward this pursuit by trying to develop newer and more sensitive methods of virus detection. It has also made attempts to understand the overall effect of the virus on the human proteome and in turn the human body. Both discovery and targeted proteomic approaches have been utilized for screening and validating the viral peptides in clinical samples to help in the diagnosis of active infection. Here, we present some of the recent/ongoing TP studies on SARS-CoV-2 virus and COVID-19.

Researchers have used TP approaches like MRM, PRM, and SWATH-DIA for developing rapid and highly sensitive methods for the diagnosis of COVID-19. One such method has been developed by Singh et al. using MRM approach, which could detect two viral peptides, AIVSTIQRKYK and QIAPGQTGK, from proteins Replicase polyprotein 1 ab and structural spike glycoprotein, respectively, in a 2.3-minute gradient with a specificity of 100% and sensitivity of 90% (Singh et al. 2020). Initially, they identified 22 peptides from four SARS-CoV-2 proteins with a 1% false discovery rate (FDR). Out of these 22 peptides, 8 peptides from three proteins (Replicase 1 ab, spike glycoprotein, and nucleoprotein) that were unique and unmodified were selected. A short MRM assay was developed using two of these peptides, AIVSTIQRKYK and QIAPGQTGK. Around 103 naso-oropharyngeal samples (including 20 control samples) were analyzed using this method and based on the results obtained, the method has a sensitivity of 90.4% and specificity of 100% to RT-PCR positive samples and true controls. A schematic of this work is shown in Figure 4.3a.

Another group has proposed an assay for the detection of specific COVID-19 peptides of nucleoprotein in highly diluted gargle solutions. The method was developed by a PRM-based targeted acquisition using high-resolution mass spectrometry (Christian Ihling et al. 2020). Additionally, a highly sensitive method has been reported recently that involves PRM-based assay for the detection of specific SARS-CoV-2 proteins in biological samples. Using Orbitrap-based high-resolution mass spectrometry, first a spectral library was generated having seven unique viral peptides belonging to two different proteins: nucleoprotein and spike glycoprotein. The two best peptides were selected for developing the final assay. The developed assay is reported as highly sensitive (with a limit of quantitation of ~390 attomoles with a limit of detection of ~200 attomoles) and a specific method and claimed to be an alternative approach to RT-PCR (Cazares et al. 2020). A simplified representation of the same is shown in Figure 4.3b. Similar kind of assay has been developed using PRM approach for the detection of two nucleoprotein peptides (GFYAQGSR and ADETQALPQR) in COVID-19 clinical samples (Gouveia, Miotello et al. 2020).

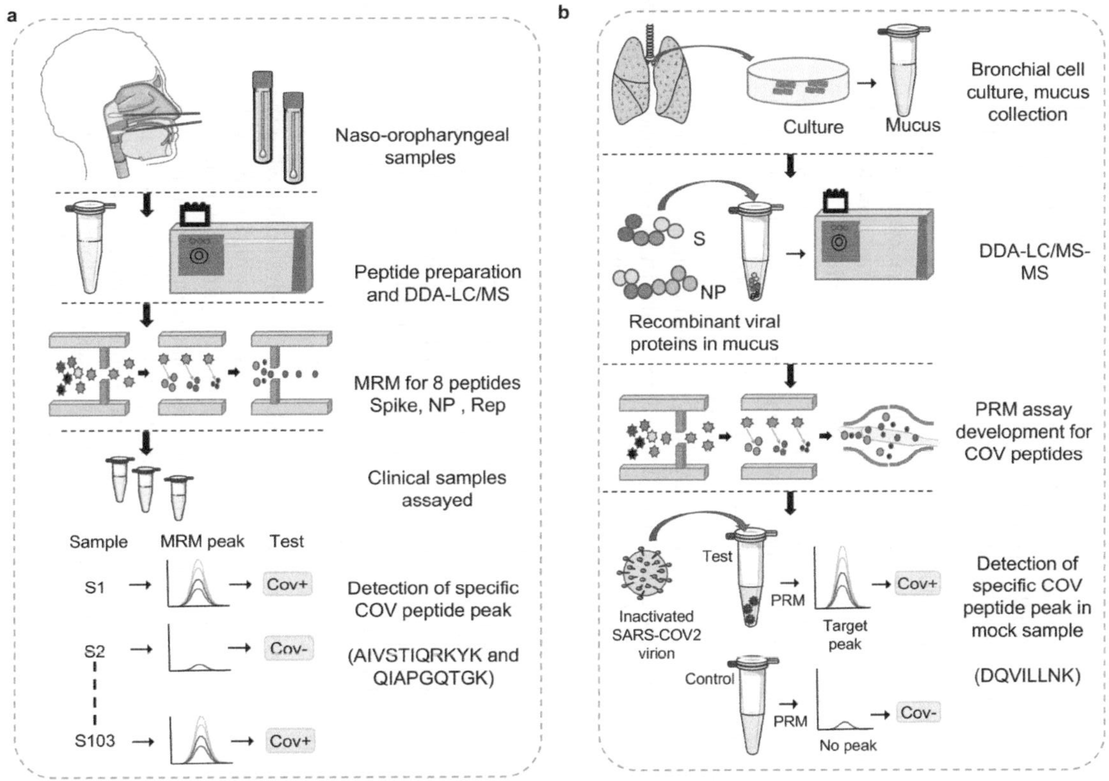

FIGURE 4.3 Overview of major studies in literature for SARS-CoV-2. (a) Representation of the work done by Singh et al. (2020) to develop a scheduled MRM assay for detecting target SARS-CoV-2 peptide in oro-nasopharyngeal samples. Viral proteins were identified using data-dependent acquisition (DDA) and eight peptides from three viral proteins (spike, nucleoprotein, and Replicase) were taken for MRM assay development. Finally, 103 clinical samples were tested against this assay with 100% specificity. (b) Representation of work done by Cazares et al. (2020) for developing a PRM-based method for the detection of SARS-CoV-2 in clinical samples. Full-length recombinant SARS-CoV-2 spike glycoprotein (S) and nucleocapsid protein (NP) were digested in an *in vitro* derived human bronchial epithelial mucus background followed by peptide analysis by DDA. Identified unique viral peptides were taken for PRM method development. Consistent and the best-performing peptides were selected for the final application. The mock test sample was prepared by spiking-in of inactivated SARS-CoV-2 virions in mucus and the developed PRM method was used to detect viral peptides in the sample.

Moreover, the combined use of MRM and PRM techniques has also been employed for developing an assay for SARS-CoV-2 detection. Cardozo et al. developed an assay using these techniques that could detect two peptides DGIIWVATEGALNTPK and IGMEVTPSGTWLTYTGAIK from the nucleoprotein of SARS-CoV-2 virus (Cardozo et al. 2020). An initial PRM run explored 17 peptides belonging to four different proteins. Out of these, nine nucleoprotein peptides were found to be most intense and unique, and thus were selected for further analysis. The two most hydrophobic peptides DGIIWVATEGALNTPK and IGMEVTPSGTWLTYTGAIK provided the best signal intensities, and hence were selected for the development of the assay. They eluted within a 2.5-minute gradient (TFC-MS system), enabling multiplexing of four samples within 10 minutes. Therefore, the system could process more than 500 samples in 24 hours. They validated the assay both qualitatively and quantitatively, using 985 samples (naso-oropharyngeal swabs) previously analyzed by real-time RT-PCR, and detected up to 84% of the positive cases with up to 97% specificity.

In search for robust diagnostic methods, scientists are also using targeted proteomics techniques along with other known techniques such as immunoaffinity purification. Renuse et al. developed an automated antibody capture-based workflow coupled to targeted high-field asymmetric ion mobility spectrometry (FAIMS)-PRM assay for the detection of SARS-CoV-2 viral peptides in nasopharyngeal swab samples (Renuse et al. 2020). The method could detect two peptides AYNVTQAFGR and QQTVTLLPAADLDDFSK, both belonging to the nucleoprotein. Results obtained from discovery experiments using nasopharyngeal swab samples from COVID-19 positive patients, recombinant viral protein and purified virus, helped them in choosing four peptides unique to the SARS-CoV-2 nucleoprotein for targeted assay development. The final assay was developed using the peptides AYNVTQAFGR and QQTVTLLPAADLDDFSK. They created an ensemble machine learning–based model for determining COVID-19 positive samples using fragment ion intensities in the PRM data after the analysis of 187 samples (116 positive and 71 negative). The same model was applied to an independent validation dataset of 176 samples (88 positive and 88 negative samples). This resulted in 97.8% sensitivity and 100% specificity that is comparative to RT-PCR tests.

SARS-CoV-2 research has also been performed using the virus cultured in the laboratory. A diagnostic method has been reported recently, which involves the orthogonal detection of a SARS-CoV-2-specific peptide of nucleoprotein for testing the viral infection with a very low limit of detection (Bezstarosti et al. 2020). Initially, for method development, the experiment was performed with research samples in which the SARS-CoV-2 virus was propagated in Vero E6 cells (African green monkey cell lines). This virus plus cell (VC) mix was taken for initial shotgun proteomics to acquire the information on all viral proteins and peptides present and it was found that of the total proteome ~4–5% proteins correspond to the virus of which more than 88% was nucleoprotein (NCAP). For the PRM experiment, a serial dilution of VC mix from 50 ng to 20 pg was prepared and NP peptides were monitored and one NP peptide AYNVTQAFGR was found to show consistent and confident results. Further, the method was extrapolated for detecting viral peptides in *in vitro* derived mucus samples spiked in with SARS-CoV-2 stock. The same method was also found to be successful in identifying the viral peptides in clinical samples, including swab and sputum samples. Also, the group used a synthetic peptide of the same sequence with a heavy isotope labeling and chromatograms were compared with viral peptides to confirm the presence of the virus and validate the method.

Another PRM-based method has been developed using laboratory-generated viral samples and applied for clinical samples in which the research group first synthesized 113 heavy labeled synthetic peptides corresponding to all possible tryptic peptides of 11 SARS-CoV-2 proteins. A spectral library was created using DDA and PRM-MS data of the synthetic peptide pool covering 98 synthetic peptides. Then supernatant from Vero E6 cells infected with SARS-CoV-2 was spiked in with synthetic peptides followed by PRM analysis in which 57 out of the targeted 98 peptides were detected, which corresponds to five viral proteins (NCAP—nucleoprotein, VME1—membrane protein, SPIKE—spike glycoprotein, ORF9B—protein 9b, and NS8—non-structural protein 8). For the final assay, 23 peptides were selected and it was tested against two diagnostic cohorts of 91 COVID-19 suspects taking their respiratory samples, in which nucleoprotein was abundantly detected (Zecha et al. 2020). Shen et al. and his group performed the combined proteomic and metabolomic analysis of 46 COVID-19 individuals to characterize the proteome and metabolome in serum of COVID-19-infected patients for identification of blood biomarkers for infection severity. A machine learning–based model was trained using 29 severe and non-severe patients and the test set was diagnosed with 70% accuracy. MRM-based targeted approach was used to validate the 22 proteins used in machine learning models as classifiers for COVID-19 positive and negative individuals using 19 infected sera samples (Shen et al. 2020). One group has recently reported an *in silico* method for the selection of unique SARS-CoV-2 peptides for diagnostic purposes. Initially, they took all possible tryptic peptides (496 peptides), followed by sequential removal of the peptides of human origin or those having similarity with other microbiomes and those from highly variable regions (Orsburn et al. 2020). Finally, they reported 24 putative viral peptides that can be utilized for any targeted or diagnostic assay as well as vaccine development. Development of more such *in silico* methods will be beneficial for not just COVID-19 diagnosis but will also help in the rapid development of MS-based diagnostics for other infectious diseases.

Mass spectrometry has also helped in generating high-throughput data for understanding the pathogenesis and severity caused by SARS-CoV-2 infection in the host body. SWATH-DIA-based mass spectrometric analysis performed on a cohort of SARS-CoV-2 patients generated comprehensive proteomic data that could identify a large panel of differentially expressed host proteins. Majorly affected proteins belong to the coagulation system, complement factors, pro-inflammatory factors, and inflammatory modulators (Messner et al. 2020). Recently, a proteomics study has reported an extensive analysis of nasopharyngeal swab samples of COVID-19 patients to study the effects of viral infection on the host at the molecular level. Several differentially expressed host proteins were identified (l-lactate dehydrogenase, interleukin-6, C-reactive protein, aspartate aminotransferase, and ferritin), the expression of which was further validated using MRM-based targeted approach (Bankar et al. 2021). A similar study was also performed on plasma samples of COVID-19 individuals to explore the proteomic alterations in response to viral infection. MRM-based approach was used for quantitative validation of differentially expressed host proteins. Few proteins like SERPINA3, AGT, FGG, APOB, and SEPRING1 were reported to be overexpressed in COVID-19 positive cases

TABLE 4.1

Summary of Peptides that Have Been Proposed for Viral Detection from Various Biological Samples

Protein	Peptides	Sample Type	Reference
Nucleoprotein	ADETQALPQR GFYAQGSR	Nasopharyngeal swab	Gouveia, Grenga et al. 2020
	DGIIWVATEGALNTPK	Naso-oropharyngeal swab	Cardozo et al. 2020
	IGMEVTPSGTWLTYTGAIK		
	RPQGLPNNTASWFTALTQHGK	Gargle solution	Christian Ihling et al. 2020
	AYNVTQAFGR QQTVTLLPAADLDDFSK	Nasopharyngeal swab	Renuse et al. 2020
	DQVILLNK	Virus-spiked mucus	Cazares et al. 2020
Spike glycoprotein	QIAPGQTGK	Naso-oropharyngeal swabs	Singh et al. 2020
	FQTLLALHR	Virus-spiked mucus	Cazares et al. 2020
Replicase polyprotein 1 ab	QIAPGQTGK	Naso-oropharyngeal swabs	Singh et al. 2020

as per their discovery and targeted proteomics data (Suvarna et al. 2021). Table 4.1 summarizes studies concerning the peptides being studied and chosen. All these studies differ in the approach and targets that were finalized but give assurance that with proper testing and optimizations, mass spectrometry–based TP assays will soon be mainstream for diagnosis of COVID-19 infections.

4.4 Conclusions and future perspectives

Numerous researchers have used TP techniques like SRM and PRM to develop several diagnostic methods for the effective detection of the coronavirus. They have identified viral peptides from clinical samples (nasopharyngeal, oropharyngeal, mucous, saliva) that can be used for diagnostic purposes. The identified viral peptides mostly belong to nucleoproteins, spike glycoproteins, or the Replicase polyprotein 1 ab of the SARS-CoV-2 virus. Some studies have also shown how their methods are more sensitive and specific than RT-PCR tests or rapid antigen tests that are currently being used (Singh et al. 2020). Many proponents point out that TP-based diagnostic methods can prove to be more economically feasible and can also be used for high-throughput testing, which is the need of the hour. Therefore, it would be interesting to see if, in the near future, these peptides (and proteins) can be used to develop diagnostic kits for testing the SARS-CoV-2 virus. Until now, most of the researchers have used TPs only to develop methods for detecting the virus. However, it must be noted that it can also be used to study the pathogenicity of the virus and explore the biology of the host response to the infection (Suvarna et al. 2021; Bankar et al. 2021). During this pandemic, targeted proteomics has served as a complementary to other standard diagnostic and research tools.

REFERENCES

Addona, Terri A., Susan E. Abbatiello, Birgit Schilling, Steven J. Skates, D. R. Mani, David M. Bunk, Clifford H. Spiegelman, et al. 2009. "Multi-Site Assessment of the Precision and Reproducibility of Multiple Reaction Monitoring-Based Measurements of Proteins in Plasma." *Nature Biotechnology* 27 (7): 633–41. https://doi.org/10.1038/nbt.1546.

Bankar, Renuka, Kruthi Suvarna, Saicharan Ghantasala, Om Shrivastav, and Jayanthi Shastri. 2021. "Proteomic Investigation Reveals Dominant Alterations of Neutrophil Degranulation and MRNA Translation Pathways in Patients with COVID-19." *ISCIENCE* 24: 102135. https://doi.org/10.1016/j.isci.2021.102135.

Bell, A. W., E. W. Deutsch, C. E. Au, R. E. Kearney, R. Beavis, S. Sechi, T. Nilsson, J. J. Bergeron, "HUPO Test Sample Working Group", et al. 2009. "A HUPO Test Sample Study Reveals Common Problems in Mass Spectrometry-Based Proteomics." *Nature Methods*. June. 6 (6): 423–430. https://doi.org/10.1038/nmeth.1333.

Bezstarosti, Karel, Mart M. Lamers, Bart L. Haagmans, and Jeroen A. A. Demmers. 2020. "Targeted Proteomics for the Detection of Sars-cov-2 Proteins." *bioRxiv*. https://doi.org/10.1101/2020.04.23.057810.

Bruderer, R., O. M. Bernhardt, T. Gandhi, S. M. Miladinović, L. Y. Cheng, S. Messner, T. Ehrenberger, V. Zanotelli, et al. 2015. "Extending the Limits of Quantitative Proteome Profiling with Data-Independent Acquisition and Application to Acetaminophen-Treated Three-Dimensional Liver Microtissues." *Molecular and Cellular Proteomics*. May. 14 (5): 1400–1410. https://doi.org/10.1074/mcp.M114.044305.

Cardozo, Karina Helena Morais, Adriana Lebkuchen, Guilherme Gonçalves Okai, Rodrigo Andrade Schuch, Luciana Godoy Viana, Aline Nogueira Olive, Carolina dos Santos Lazari, et al. 2020. "Establishing a Mass Spectrometry-Based System for Rapid Detection of SARS-CoV-2 in Large Clinical Sample Cohorts." *Nature Communications* 11 (1). https://doi.org/10.1038/s41467-020-19925-0.

Cazares, Lisa H., Raghothama Chaerkady, Shao Huan Samuel Weng, Chelsea C. Boo, Raffaello Cimbro, Hsiang En Hsu, Sarav Rajan, et al. 2020. "Development of a Parallel Reaction Monitoring Mass Spectrometry Assay for the Detection of SARS-CoV-2 Spike Glycoprotein and Nucleoprotein." *Analytical Chemistry* 92 (20): 13813–21. https://doi.org/10.1021/acs.analchem.0c02288.

Chaimayo, Chutikarn, Bualan Kaewnaphan, Nattaya Tanlieng, Niracha Athipanyasilp, Rujipas Sirijatuphat, Methee Chayakulkeeree, Nasikarn Angkasekwinai, et al. 2020. "Rapid SARS-CoV-2 Antigen Detection Assay in Comparison with Real-Time RT-PCR Assay for Laboratory Diagnosis of COVID-19 in Thailand." *Virology Journal* 17 (1). https://doi.org/10.1186/s12985-020-01452-5.

de Graaf, Erik L., A. F. Maarten Altelaar, Bas van Breukelen, Shabaz Mohammed, and Albert J. R. Heck et al. 2011. "Improving SRM Assay Development: A Global Comparison between Triple Quadrupole, Ion Trap, and Higher Energy CID Peptide Fragmentation Spectra." *Journal of Proteome Research* 10 (9), 4334–4341. https://doi.org/10.1021/pr200156b.

Desiderio, Dominic M., and Masaaki Kai. 1983. "Preparation of Stable Isotope-Incorporated Peptide Internal Standards for Field Desorption Mass Spectrometry Quantification of Peptides in Biologic Tissue." *Biological Mass Spectrometry* 10 (8): 471–79. https://doi.org/10.1002/bms.1200100806.

Dramé, Moustapha, Maturin Tabue Teguo, Emeline Proye, Fanny Hequet, Maxime Hentzien, Lukshe Kanagaratnam, and Lidvine Godaert. 2020. "Should RT-PCR Be Considered a Gold Standard in the Diagnosis of COVID-19?" *Journal of Medical Virology* 93 (1). John Wiley and Sons Inc: 137–38. https://doi.org/10.1002/jmv.25996.

Elslande, J. van, E. Houben, M. Depypere, A. Brackenier, S. Desmet, E. André, M. van Ranst, K. Lagrou, and P. Vermeersch. 2020. "Diagnostic Performance of Seven Rapid IgG/IgM Antibody Tests and the Euroimmun IgA/IgG ELISA in COVID-19 Patients." *Clinical Microbiology and Infection* 26 (8): 1082–87. https://doi.org/10.1016/j.cmi.2020.05.023.

Fenn, John B., Matthias Mann, Chin Kai Meng, Shek Fu Wong, and Craig M. Whitehouse. 1989. "Electrospray Ionization for Mass Spectrometry of Large Biomolecules." *Science* 246 (4926). American Association for the Advancement of Science: 64–71. https://doi.org/10.1126/science.2675315.

Gouveia, Duarte, Lucia Grenga, Jean-Charles Gaillard, Fabrice Gallais, Laurent Bellanger, Olivier Pible, and Jean Armengaud. 2020. "Shortlisting SARS-CoV-2 Peptides for Targeted Studies from Experimental Data-Dependent Acquisition Tandem Mass Spectrometry Data." *PROTEOMICS*, June. https://doi.org/10.1002/pmic.202000107.

Gouveia, Duarte, Guylaine Miotello, Fabrice Gallais, Jean Charles Gaillard, Stéphanie Debroas, Laurent Bellanger, Jean Philippe Lavigne, et al. 2020. "Proteotyping SARS-CoV-2 Virus from Nasopharyngeal Swabs: A Proof-of-Concept Focused on a 3 Min Mass Spectrometry Window." *Journal of Proteome Research* 19 (11): 4407–16. https://doi.org/10.1021/acs.jproteome.0c00535.

Grossegesse, Marica, Felix Hartkopf, Andreas Nitsche, Lars Schaade, Joerg Doellinger, and Thilo Muth. 2020. "Perspective on Proteomics for Virus Detection in Clinical Samples." *Journal of Proteome Research* 19 (11). American Chemical Society: 4380–88. https://doi.org/10.1021/acs.jproteome.0c00674.

Henderson, Clark M., James G. Bollinger, Jessica O. Becker, Jennifer M. Wallace, Thomas J. Laha, Michael J. MacCoss, and Andrew N. Hoofnagle. 2017. "Quantification by Nano Liquid Chromatography Parallel Reaction Monitoring Mass Spectrometry of Human Apolipoprotein A-I, Apolipoprotein B, and Hemoglobin A1c in Dried Blood Spots." *PROTEOMICS—Clinical Applications* 11 (7–8): 1600103. https://doi.org/10.1002/prca.201600103.

Hoofnagle, Andrew N., Jessica O. Becker, Mark H. Wener, and Jay W. Heinecke. 2008. "Quantification of Thyroglobulin, a Low-Abundance Serum Protein, by Immunoaffinity Peptide Enrichment and Tandem Mass Spectrometry." *Clinical Chemistry* 54 (11): 1796–804. https://doi.org/10.1373/clinchem.2008.109652.

Huillet, Céline, Annie Adrait, Dorothée Lebert, Guillaume Picard, Mathieu Trauchessec, Mathilde Louwagie, Alain Dupuis, et al. 2012. "Accurate Quantification of Cardiovascular Biomarkers in Serum Using Protein Standard Absolute Quantification (PSAQ™) and Selected Reaction Monitoring." *Molecular and Cellular Proteomics* 11 (2): M111.008235. https://doi.org/10.1074/mcp.M111.008235.

Ihling, Christian, Dirk Tänzler, Sven Hagemann, Astrid Kehlen, Stefan Hüttelmaier, Christian Arlt, and Andrea Sinz. 2020. "Mass Spectrometric Identification of SARS-CoV-2 Proteins from Gargle Solution Samples of COVID-19 Patients." *Journal of Proteome Research* 19 (11): 4389–92. https://doi.org/10.1021/acs.jproteome.0c00280.

Kasteren, Puck B. van, Bas van der Veer, Sharon van den Brink, Lisa Wijsman, Jørgen de Jonge, Annemarie van den Brandt, Richard Molenkamp, Chantal B. E. M. Reusken, and Adam Meijer. 2020. "Comparison of Seven Commercial RT-PCR Diagnostic Kits for COVID-19." *Journal of Clinical Virology* 128 (July). https://doi.org/10.1016/j.jcv.2020.104412.

Kennedy, Jacob J., Jeffrey R. Whiteaker, Laura C. Kennedy, Dustin E. Bosch, Melissa L. Lerch, Regine M. Schoenherr, Lei Zhao, et al. 2021. "Quantification of Human Epidermal Growth Factor Receptor 2 by Immunopeptide Enrichment and Targeted Mass Spectrometry in Formalin-Fixed Paraffin-Embedded and Frozen Breast Cancer Tissues." *Clinical Chemistry* 67 (7): 1008–18. https://doi.org/10.1093/clinchem/hvab047.

Kopylov, A. T., V. G. Zgoda, A. V. Lisitsa, A. I. Archakov. 2013. "Combined Use of Irreversible Binding and MRM Technology for Low- and Ultralow Copy-Number Protein Detection and Quantitation." *Proteomics*, March. 13 (5): 727–42. https://doi.org/10.1002/pmic.201100460.

MacLean, Brendan, Daniela M. Tomazela, Nicholas Shulman, Matthew Chambers, Gregory L. Finney, Barbara Frewen, Randall Kern, David L. Tabb, Daniel C. Liebler, and Michael J. MacCoss. 2010. "Skyline: An Open Source Document Editor for Creating and Analyzing Targeted Proteomics Experiments." *Bioinformatics* 26 (7): 966–68. https://doi.org/10.1093/bioinformatics/btq054.

Marx, Vivien. 2013. "Targeted Proteomics." *Nature Methods* 10 (1): 19–22. https://doi.org/10.1038/nmeth.2285.

Messner, Christoph B., Vadim Demichev, Daniel Wendisch, Laura Michalick, Matthew White, Anja Freiwald, Kathrin Textoris-Taube, et al. 2020. "Ultra-High-Throughput Clinical Proteomics Reveals Classifiers of COVID-19 Infection." *Cell Systems* 11 (1): 11–24.e4. https://doi.org/10.1016/j.cels.2020.05.012.

Orsburn, Benjamin, Conor Jenkins, Sierra M. Miller, Benjamin A. Neely, and Namandje M. Bumpus. 2020. "In Silico Approach Toward the Identification of Unique Peptides from Viral Protein Infection: Application to COVID-19." *SSRN Electronic Journal*, May. https://doi.org/10.2139/ssrn.3589835.

Park, Joonho, Hyeon Jeong Oh, Dohyun Han, Joseph I. Wang, In Ae Park, Han Suk Ryu, and Youngsoo Kim. 2020. "Parallel Reaction Monitoring-Mass Spectrometry (PRM-MS)-Based Targeted Proteomic Surrogates for Intrinsic Subtypes in Breast Cancer: Comparative Analysis with Immunohistochemical Phenotypes." *Journal of Proteome Research* 19 (7): 2643–53. https://doi.org/10.1021/acs.jproteome.9b00490.

Pino, Lindsay K., Brian C. Searle, James G. Bollinger, Brook Nunn, Brendan MacLean, and Michael J. MacCoss. 2020. "The Skyline Ecosystem: Informatics for Quantitative Mass Spectrometry Proteomics." *Mass Spectrometry Reviews* 39 (3). John Wiley and Sons Inc: 229–44. https://doi.org/10.1002/mas.21540.

Reiter L, O. Rinner, P. Picotti, R. Hüttenhain, M. Beck, M. Y. Brusniak, M. O. Hengartner, R. Aebersold, et al. 2011. "Automated Data Processing and Statistical Validation for Large-Scale SRM Experiments." *Nature Methods*. May. 8(5): 430–435. https://doi.org/10.1038/nmeth.1584.

Ren, Xiaoshuai, Yan Liu, Hongtao Chen, Wei Liu, Zhaowang Guo, Yaqin Zhang, Chaoqun Chen, et al. 2020. "Application and Optimization of RT-PCR in Diagnosis of SARS-CoV-2 Infection." *SSRN Electronic Journal*, March. https://doi.org/10.2139/ssrn.3546086.

Renuse, Santosh, Patrick M. Vanderboom, Anthony D. Maus, Jennifer V. Kemp, Kari M. Gurtner, Anil K. Madugundu, Sandip Chavan, et al. 2020. "Development of Mass Spectrometry-Based Targeted Assay for Direct Detection of Novel SARS-CoV-2 Coronavirus from Clinical Specimens." *MedRxiv*. https://doi.org/10.1101/2020.08.05.20168948.

Shen, Bo, Xiao Yi, Yaoting Sun, Xiaojie Bi, Juping Du, Chao Zhang, Sheng Quan, et al. 2020. "Proteomic and Metabolomic Characterization of COVID-19 Patient Sera." *Cell* 182 (1): 59–72.e15. https://doi.org/10.1016/j.cell.2020.05.032.

Sherman, Jamie, Matthew J. McKay, Keith Ashman, and Mark P. Molloy. 2009. "How Specific Is My SRM?: The Issue of Precursor and Product Ion Redundancy." *PROTEOMICS* 9 (5): 1120–23. https://doi.org/10.1002/pmic.200800577.

Singh, Praveen, Rahul Chakraborty, Robin Marwal, V. S. Radhakrishan, Akash Kumar Bhaskar, Himanshu Vashisht, Mahesh S. Dhar, et al. 2020. "A Rapid and Sensitive Method to Detect SARS-CoV-2 Virus Using Targeted-Mass Spectrometry." *Journal of Proteins and Proteomics* 11 (3): 159–65. https://doi.org/10.1007/s42485-020-00044-9.

Suvarna, Kruthi, Deeptarup Biswas, Medha Gayathri J. Pai, Arup Acharjee, Renuka Bankar, Viswanthram Palanivel, Akanksha Salkar, et al. 2021. "Proteomics and Machine Learning Approaches Reveal a Set of Prognostic Markers for COVID-19 Severity with Drug Repurposing Potential." *Frontiers in Physiology* 12 (April): 432. https://doi.org/10.3389/FPHYS.2021.652799.

Tabb, D. L., L. Vega-Montoto, P. A. Rudnick, A. M. Variyath, A. J. Ham, D. M. Bunk, L. E. Kilpatrick, D. D. Billheimer, R. K. Blackman, H. L. Cardasis, et al. 2010. "Repeatability and Reproducibility in Proteomic Identifications by Liquid Chromatography-Tandem Mass Spectrometry." *Journal of Proteome Research,* February 5. 9 (2): 761–776. https://doi.org/10.1021/pr9006365.

Vandenberg, Olivier, Delphine Martiny, Olivier Rochas, Alex van Belkum, and Zisis Kozlakidis. 2020. "Considerations for Diagnostic COVID-19 Tests." *Nature Reviews Microbiology* 19 (3). Nature Research: 171–83. https://doi.org/10.1038/s41579-020-00461-z.

Worboys, Jonathan D., John Sinclair, Yinyin Yuan, and Claus Jørgensen. 2014. "Systematic Evaluation of Quantotypic Peptides for Targeted Analysis of the Human Kinome." *Nature Methods* 11 (10): 1041–44. https://doi.org/10.1038/nmeth.3072.

Zecha, Jana, Chien Yun Lee, Florian P. Bayer, Chen Meng, Vincent Grass, Johannes Zerweck, Karsten Schnatbaum, et al. 2020. "Data, Reagents, Assays and Merits of Proteomics for SARS-CoV-2 Research and Testing." *Molecular and Cellular Proteomics* 19 (9): 1503–22. https://doi.org/10.1074/mcp.RA120.002164.

Zhang, Haixia, Qinfeng Liu, Lisa J. Zimmerman, Amy Joan L. Ham, Robbert J. C. Slebos, Jamshedur Rahman, Takefume Kikuchi, et al. 2011. "Methods for Peptide and Protein Quantitation by Liquid Chromatography- Multiple Reaction Monitoring Mass Spectrometry." *Molecular and Cellular Proteomics* 10 (6): M110.006593. https://doi.org/10.1074/mcp.M110.006593.

Zhu, Na, Dingyu Zhang, Wenling Wang, Xingwang Li, Bo Yang, Jingdong Song, Xiang Zhao, et al. 2020. "A Novel Coronavirus from Patients with Pneumonia in China, 2019." *New England Journal of Medicine* 382 (8): 727–33. https://doi.org/10.1056/nejmoa2001017.

5

Metabolomics

Role in pathobiology and therapeutics of COVID-19

Shalini Aggarwal, Nirjhar Banerjee, Shashwati Parihari, Jyotirmoy Roy, Kharanshu Bojak, and Rhythm Shah

CONTENTS

List of abbreviations

BAL – Bronchoalveolar lavage
CV – Coefficient of variation
EWAS/ExWAS – Exposome-wide association studies
FBS – Fetal bovine serum
FTIR – Fourier transform infrared spectroscopy
GC-MS – Gas chromatography MS
HBSS – Hank balanced salt solution
HILIC – Hydrophilic interaction liquid chromatography
HMDB – Human Metabolome Database
HPLC-MS – High-performance liquid chromatography MS

DOI: 10.1201/9781003220787-5

IS – Internal standards
KEGG – Kyoto Encyclopedia of genes and genomes
LAF – Laminar airflow
LC-(FTIR)
LC – Liquid chromatography
MS – Mass spectrometry
NMR – Nuclear magnetic resonance spectrometry
PBS-(VTM)
PBS – Phosphate buffer saline
QC – Quality control
RP – Reverse-phase
SDS – Sodium dodecyl sulfate
UV – Ultraviolet
VTM – Viral transport media

5.1 Introduction

The term "metabolomics" was first coined in the late 1990s (Oliver et al. 1998). The term refers to the study of all metabolites in a particular organism or system. Metabolites are essentially small molecules of various chemical properties, notably considered to be <1,000 Da in size, which plays crucial roles in various pathways in an organism. Metabolites include amino acids, small carbohydrate moieties, nucleotides, and their components nucleosides, lipids, and other biomolecules, involved in all metabolic pathways throughout the central dogma of an organism (Aggarwal et al. 2021).

Being the intermediates and by-products of several metabolic pathways, metabolites are the immediate broadcasting signals of the physiological condition of a biological system and therefore making metabolomics a direct functional report to the same (Roessner and Bowne 2009). Although metabolites can be extracted from any biological specimen such as cellular samples and extracellular biofluids, the latter is the one that gives metabolomics an edge regarding the ease of access to specimens to be tested, including blood plasma, urine, saliva, sweat, semen, feces, and also internal samples like cerebrospinal fluid, bronchoalveolar lavage, amniotic fluid, etc. Minimal sample processing, involvement of highly developed as well as always advancing high-throughput systems like nuclear magnetic resonance spectroscopy (NMR), mass spectrometry (MS), liquid chromatography (LC) along with powerful bioinformatics and statistical analytical techniques (Tsutsui et al. 2011; A. Zhang et al. 2015) add to the credibility and feasibility of performing metabolomics experiments at large scale, thereby increasing the reliability of the results (Figure 5.1a).

5.1.1 Role of metabolomics in disease pathobiology

Metabolome being the immediate broadcasting signals of the physiological condition, metabolites deem to be early indicators of any change in the organism due to diseased or pathological condition (Figure 5.1b). Metabolomics study of progressive disease samples may provide critical insights about early prognostic

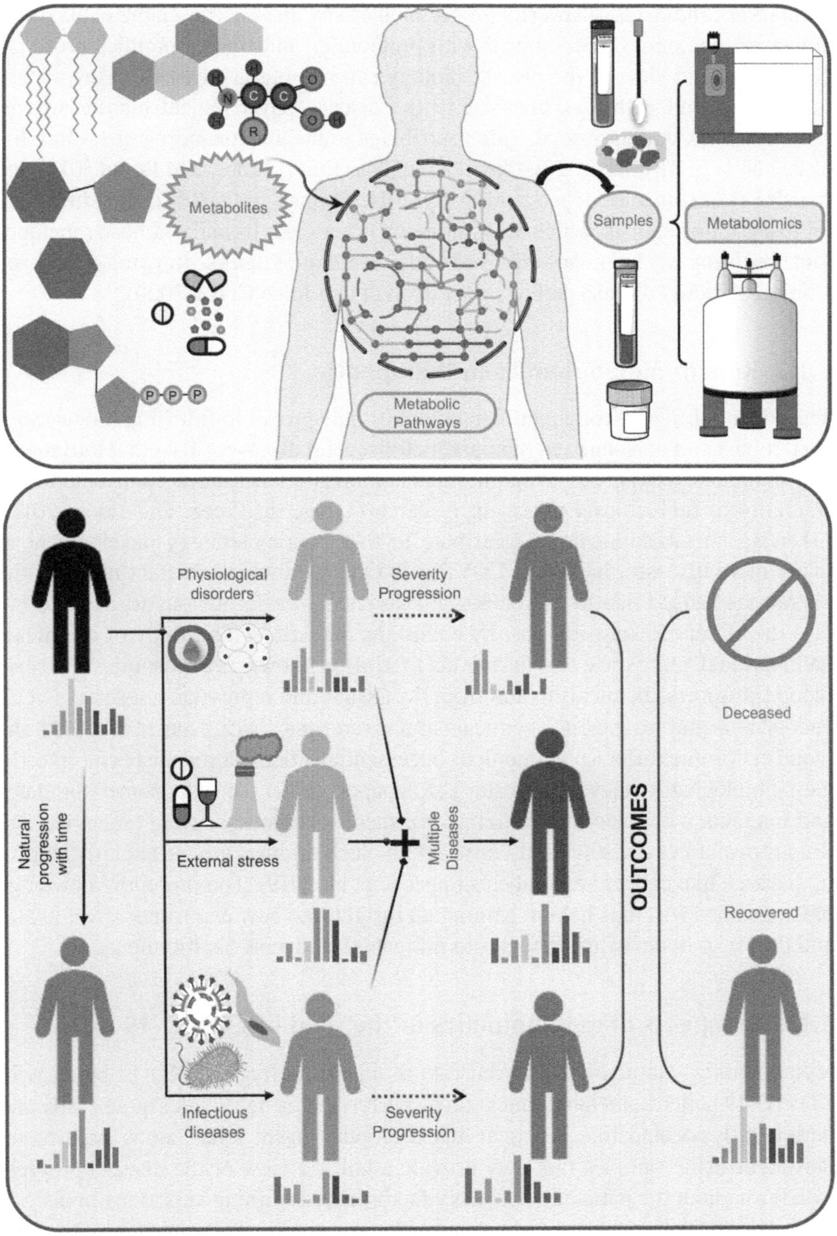

FIGURE 5.1 Metabolomics and its applications. (a) Metabolites–Metabolic pathways–metabolomics: schematic of conceptual understanding. (b) Role of metabolomics in disease diagnosis, prognosis, and therapeutics—a conceptual illustration (each human icon denotes a typical physiological condition/stage; the bar graphs below each condition/stage denote their symbolic metabolome profiles which can be further used for understanding diagnosis, prognosis, or therapeutics).

biomarker candidates of severity progression of any disease (A. Zhang et al. 2015). When metabolomics is combined with proteomics and transcriptomics profile, a clearer, holistic view of the disease biology can be obtained by exhibiting altered metabolites and pathways involved in the disease physiological manifestations. Metabolomics studies have already contributed to the understanding of several diseases (N. P. Long et al. 2020), like tuberculosis (Preez, Luies, and Loots 2017) and malaria (Lakshmanan, Rhee, and Daily 2011; Salinas et al. 2014), shedding light on disease mechanism which was unknown otherwise. In malaria hosts, metabolome profiling has led to an improved understanding of disease transmission (Joice Cordy 2020) and disease pathogenesis (Colvin and Joice Cordy 2020).

5.1.2 Role of metabolomics in therapeutic

The advent of high-throughput metabolomics may propel in fulfilling the need for even robust and personalized biomarkers for several diseases (Trivedi, Hollywood, and Goodacre 2017). For example, metabolomics studies have contributed substantially to tuberculosis treatment research (Luies, du Preez, and Loots 2017). Likewise, metabolomics can contribute to therapeutics strategy development in many more diseases, including COVID-19. Our metabolome buffers not only the alterations caused due to the pathogens we encounter in our day-to-day life but also the effect of exposures cast by environmental stress. The study of chemicals an individual gets exposed to during his lifetime is known as exposomics. Because blood transports chemicals to and from the tissues and represents a reservoir of all endogenous and exogenous chemicals at a given time (Nicholson et al. 2012), the blood exposomes offer a parsimonious but essentially unexplored means for interrogating biologically relevant exposures (Rappaport 2012). The exposome modulates and interacts with drug-metabolizing enzymes, xenobiotic-sensing receptors, and drug transporters (Pristner and Warth 2020). Such findings are particularly helpful in diseases like cancer and diabetes (Agache et al. 2019). The in-depth knowledge of exposomes will thus help in gaining an insight into how exposures affect health and thus help in devising strategies to mitigate the adverse health impact.

5.1.3 Prospects of metabolomics in the field of COVID-19

Metabolomics can also play a vital role in understanding COVID-19 biology. In COVID-19 patients, metabolomics can be performed using samples quite easily and peripherally located like nasopharyngeal/oropharyngeal swab, along with blood plasma or urine samples that may provide a holistic view of the disease progression throughout the patient body. A key factor in determining variations in disease severity of COVID-19 may be the exposome, or measure of all exposures across an individual's lifetime that consists of environmental toxins, pharmacological treatments, lifestyle, diet, and any other absorbable compound (Naughton et al. 2020). The transmission of SARS-CoV-2, the progress, and the outcome of COVID-19 for each individual depend on the virus or infection characteristics (e.g., viral load), existing comorbidities, and the multitude of exposures in various surroundings that belong to the specific external and internal human exposome domains (Jang,

Han, and Rhee 2020; Richardson et al. 2020; Wild 2012). The clinical manifestations are endogenic changes and can be studied using the various sample types and omics technologies, but the exposome also plays a crucial role in determining a host response.

The current inventory of exposomes, the small molecules, and metals associated with various diseases can be found in the Blood Exposome Database and Norman Network Database. These compounds are not found in the Human Metabolome Database (HMDB) as they are not directly produced in any metabolic pathway. However, the compounds often appear in blood, either due to the food the individual is taking or due to the drugs that are administered to the patient, or the environment in which the individual is living. These compounds, on entering the body, either affect any of the biological pathways or get degraded, and the degraded product, in turn, causes anomalies in the metabolic pathways. With the recognition of their health significance, these chemicals can be routinely monitored for clinical interventions such as cholesterol (Shi et al. 2021), folic acid, vitamins, and regulated pollutants such as lead, arsenic, benzene, and PCBs (Rappaport 2012). Even for the COVID-19 pandemic, the transmission of SARS-CoV-2, the progress, and the outcome of COVID-19 for each individual depend on the virus or infection characteristics (e.g., viral load), existing comorbidities, and the multitude of exposures in various settings such as school, home, and work that belong to the specific external and internal human exposome domains (Jang, Han, and Rhee 2020; Richardson et al. 2020; Wild 2012). Therefore, the application of a comprehensive framework of the metabolome with the exposome profile (Figure 5.1b) may lead to effective evaluation of public health measures.

5.2 Overview workflow of metabolomics experiments

5.2.1 Inactivation strategies for viral load

Viral inactivation refers to the irreversible inactivation of any potential virus present in the sample which is essential to reduce the risk of viral infection during sample preparation (Auerswald et al. 2021). Viral inactivation of SARS-CoV-2 can be achieved by changing the environment of the virus where it resides, through chemical (e.g., detergent, alcohol, altering pH, formalin treatment) and physical (e.g., heat, ultraviolet (UV) rays) alteration (Hemati et al. 2020; Welch et al. 2020; Darnell et al. 2004; Heßling et al. 2020; Hessling et al. 2020; Pastorino et al. 2020; Martins et al. 2020; Welch et al. 2020; Shen et al. 2020; Patterson et al. 2020; Locher et al. 2020; Rodino et al. 2020; Rogers et al. 2020) (Figure 5.2a).

5.2.2 Sample preparation for metabolome profiling

The overall workflow for metabolomics analysis includes sample preparation, compound separation, compound identification, statistical data analysis, and data interpretation (Figure 5.2b). The sample must be stored at −80°C to avoid

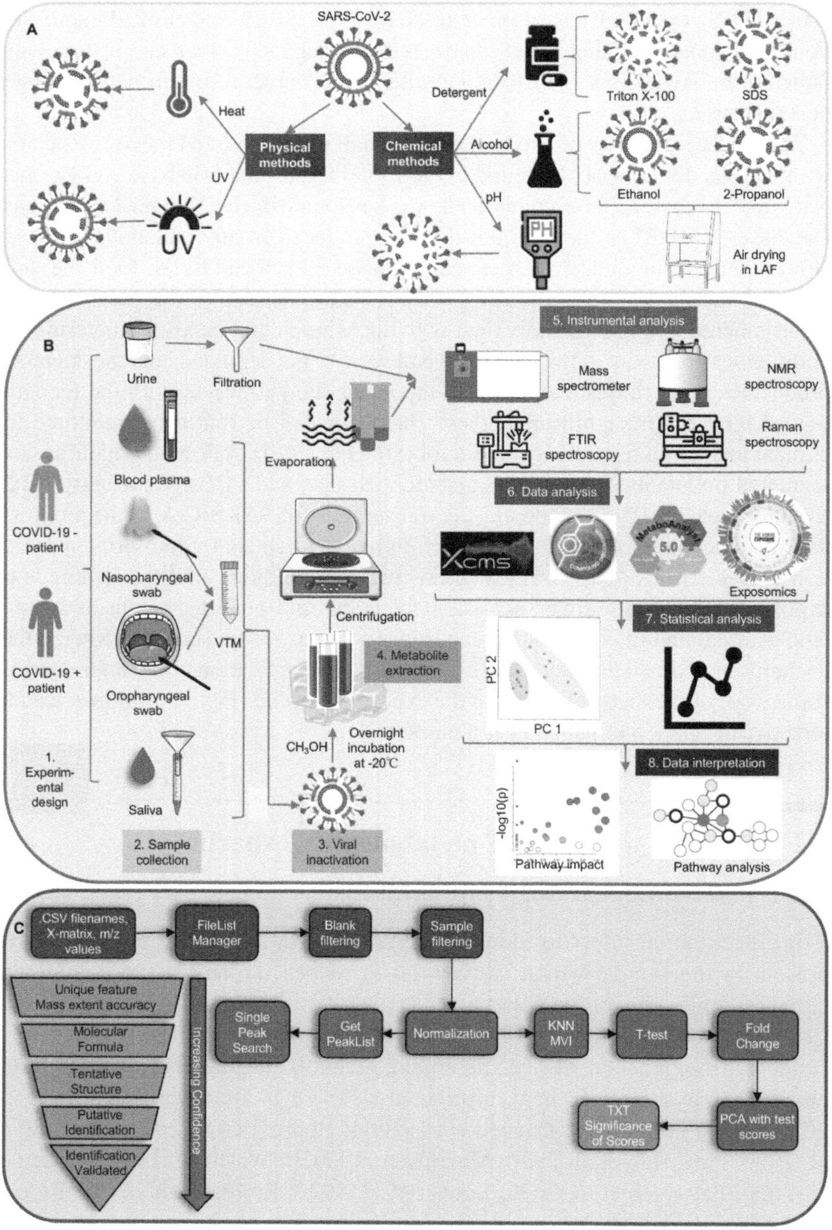

FIGURE 5.2 Schematic workflow of metabolomics experiment. (a) Physical and chemical strategies of viral inactivation before sample preparation. (b) Schematic workflow of metabolomics analysis showing experimental design, sample collection, metabolite extraction, and instrumental analysis followed by data analysis and data interpretation. (c) Detailed pipeline for metabolites identification, annotation, and data analysis.

degradation of metabolites (Chetwynd, Dunn, and Rodriguez-Blanco 2017). Sample type selection and sample preparation play a crucial role in metabolome profiling (Gong et al. 2017). Metabolite extraction can be done from various biospecimen such as blood plasma, serum, urine, nasopharyngeal swab, oropharyngeal swab, saliva, broncho-alveolar lavage (BAL) fluid, sputum, tissue sample, and more (Costa dos Santos Junior et al. 2020; Shen et al. 2020; Bernatchez and McCall 2020; Maras et al. 2020). Metabolite extraction being the major bottleneck of untargeted metabolome profiling aims to effectively extract enriched metabolites without other interfering agents such as high molecular weight proteins and salts (Álvarez-Sánchez, Priego-Capote, and Luque de Castro, 2010). Direct injection of biofluids for metabolome analysis reduces the metabolite loss; however, the presence of salt and protein may hinder the spectral data and may give false-negative results. Therefore, it is necessary to desalt and deproteinize samples before injecting them for further analysis (Álvarez-Sánchez, Priego-Capote, and Luque de Castro, 2010). Methanol and acetonitrile can be used as desalting and deproteinizing agent during metabolite extraction (Jain, Li, and Chen 2019; Amir Hashim et al. 2020). However, the extraction steps and choice of organic solvents should be based on the biospecimen. Fluid biospecimens involve minimal steps for metabolite extraction, which includes the addition of organic solvents (e.g., methanol, ethanol, chloroform, acetonitrile, or acetone) at different temperatures for the separation of polar and non-polar components, protein precipitation, and to concentrate sample as per the requirement (Figure 5.2a) (Álvarez-Sánchez, Priego-Capote, and Luque de Castro, 2010). In the case of plasma samples, protein precipitation done by the addition of chilled methanol and supernatant is collected for the MS analysis (Shen et al. 2020). Miller et al. reported that direct injection of urine samples can be done in MS for metabolome profiling (Miller et al. 2019). However, Bernatchez et al. reported tissue/cell-based biospecimens to require homogenization of the sample before proceeding for the metabolite extraction step (Bernatchez and McCall 2020).

Extracted metabolites from various biospecimens should be stored immediately after extraction in liquid or dry form at −20°C or −80°C for future measurements (Sauerschnig et al. 2017). Identification and quantification of extracted metabolites can be analyzed by using various analytical techniques such as NMR spectroscopy and MS.

5.2.3 Instrumental setup

Recently, various analytical techniques have been evolved for metabolome profiling which includes FTIR, Raman spectroscopy, NMR spectroscopy, and MS (Figure 5.2b) (Segers et al. 2019; Mahmud and Garrett 2020; Instituto Tecnológico de Aguascalientes and Sánchez-Brito, 2020). Mass spectrometry is usually coupled with high-performance liquid chromatography (HPLC-MS) or gas chromatography (GC-MS) (Zeki et al. 2020). LC-MS is generally used to detect and identify medium polar to polar compounds, whereas GC-MS is mainly used to identify non-polar and volatile metabolites (Zeki et al. 2020). The separation of analytes depends on the column length, diameter, particle size, stationary phase, mobile phase, and column temperature. Column selection should be made based on the widest coverage for the screening of more number of unknown compounds which requires prior knowledge on the target analyte type (Gough, Bahaghighat, and Synovec 2019). The separation of analyte is mainly based on the hydrophobic and hydrophilic nature of the compound. C18 column and hydrophilic interaction liquid chromatography (HILIC) column are mainly used for the separation of hydrophobic compounds by reverse-phase (RP) and hydrophilic compounds, respectively (Schwaiger et al. 2019; Rampler et al. 2018). However, a mixed-mode column is now being used for the wide range coverage of both hydrophobic and hydrophilic metabolites. The selection of the mobile phase also plays a key role in the separation of metabolites (Maras et al. 2020; Shen et al. 2020). In RP chromatography, polar solvents (e.g., methanol, acetonitrile) are used as the mobile phase. Hydrophobic analytes are attached to the stationary phase and hydrophilic analytes pass with the mobile phase and elute first. Blank samples (i.e., all the organic solvents except biospecimen) are injected before the sample run to remove background compounds coming from organic solvents (Shen et al. 2020).

LC-MS is highly sensitive and can identify a wide range of metabolites from each sample (Zheng et al. 2020; Doğan et al. 2020). MS1 is used for the quantification of the metabolites, and MS/MS is used for the identification of the metabolites. Generally, protonated analytes are detected in the positive ion mode, and deprotonated analytes are detected in the negative ion mode (Yuan et al. 2012). However, for metabolite quantification purposes, it requires more time than the NMR setup. 1H NMR (proton NMR) mainly provides compound quantification more precisely. The spectral peak obtained for one metabolite is proportional to the concentration of the metabolites in the sample. NMR also provides structural information about the metabolites (Snytnikova et al. 2019; Beckonert et al. 2007; Costa dos Santos Junior et al. 2020). However, NMR is less sensitive, and the total number of metabolites obtained from the NMR spectroscopy is very less in comparison to the metabolites obtained from LC-MS analysis. To eliminate false discoveries, it is critical to have multiple control runs (e.g., blank, standard mixtures) and quality control (QC) check steps (e.g., QC pools, internal standard) (Godzien et al. 2015).

5.2.4 Quality control check

An experiment is not valid without a QC check. Metabolomics being highly dynamic makes it prone to errors; hence, multiple QC steps such as (1) QC pools; (2) internal standards (IS); (3) correlation plots are involved in the experimental setup to check the sample run and data quality.

a. *QC pools:* These are prepared by pooling equal volumes of all samples in a given experiment to monitor the instrumental variability and batch effects (Godzien et al. 2015). Generally, QC pools should be run in between each five-sample run. Internal standards are administered with the sample to check various sample preparation variability, instrument variability, and used as a data QC check (Hewavitharana 2011).

b. *Internal standard:* IS selection should focus on these main criteria: (1) it should not be present in the sample or does not occur naturally or synthetic isotopically labeled one; (2) it should not hinder the detection of other compounds present in the sample; (3) concentration of IS should be comparable to that of the other metabolites; (4) it should be identified in every sample; (5) it should show similar intensity in each sample (Boysen et al. 2018; Stokvis, Rosing, and Beijnen 2005). Ideally, two ISs should be used in case of sample preparation, one at the starting of the metabolite extraction to monitor sample preparation variability and the other one just before injecting sample in the instrument to check the instrument sample run variability (Stokvis, Rosing, and Beijnen 2005; Shen et al. 2020). The percentage of coefficient of variation (CV) of ISs in replicates of each sample is calculated to check the data quality. CV % of the IS should be below 20–30% in each sample and also in QC pools. Day-wise CV % should be calculated as part of the QC check to track the batch variation (Shen et al. 2020; Maras et al. 2020).

c. *Correlation plots:* A correlation matrix between the different cohorts of samples and QC pools should be plotted to determine the correlation among the samples.

d. *Blank or background subtraction:* Contaminants from chemicals (such as organic solvents used in the mobile phase during the sample run) may give false-negative results during data analysis. These chemical interferants can be removed from data by background subtraction and normalization from the blank injections during the sample run to avoid misinterpretation of the data (Caesar, Kvalheim, and Cech 2018). Background subtraction can be software-based (e.g., Compound Discoverer) or manual checking of the raw value trend of identified significant metabolites in blank samples (Cerrato et al. 2020).

Endogenous metabolites present in QC pools can also be monitored to check the experimental variability. Primary metabolomics data analysis includes metabolite identification/quantification, chromatographic peak alignment, prediction of the composition of the compound, database searching against different databases (e.g., Human Metabolome Database [HMDB], Kyoto Encyclopedia of Genes and Genomes [KEGG], BioCyc), mass spectrum visualization, and statistical analysis (Gorrochategui et al. 2016).

5.2.5 Metabolite identification and statistical analysis

Metabolomics data analysis workflow can be implemented for two widely used non-targeted metabolomics modalities: (a) to transform the small molecule data into known compounds and (b) molecular pathways. This involves compound identification, determines real differences between samples, and elucidates biological pathways with integrated, using powerful software, workflows of software such as Galaxy, Compound Discoverer, and more.

In the case of LC-MS, there is a single .mzMLraw file for each sample. As the first step of processing, another workflow tool XCMS-Basic-Birmingham-Pipe is used for the current implementation. It reads in the individual spectra and aligns the spectral features across the samples using the R-package XCMS, returning three .csv files containing the X matrix, filenames, and *m/z* values, respectively. For further processing, data is converted to DSO format using create DSO that combines the X data matrix file with row and column label information and class labels describing whether each sample is biological or blank, blank filtering that removes the samples that appear strongly in the blanks, and sample filtering that removes unreliable peaks (Figure 5.2c). Another freely accessible online tool, Mummichog, is a free Python program that enables one to organize metabolic networks to predict functional activity directly from feature tables, bypassing metabolite identification. Thus, high-quality hypotheses can be quickly generated from an LC-MS data table.

To manipulate the DSO structure, workflows such as "set include", define the "include" flag variable for either samples or *m/z*. "Add class list" appends a list of sample groupings or classifications, "get class list" extracts either the text labels or integer representation of any class list in the DSO, "get peak list" extracts the *m/z* values and average peak intensities from the DSO, "get X matrix" extracts the data (X) matrix as a .csv file, and "get axis scale" extracts the values stored in the "axis scale" variable of the DSO. To prepare the X data matrix for statistical analysis tools, normalization, missing values imputation using a KNN algorithm, and log transformation to stabilize the technical variance across all peaks are used (Figure 5.2c).

Once the pXn matrix is obtained in the form of excel or csv format, platforms like MetaboAnalyst R are used to carry out fold change tests, *t*-test, volcano plots, and also for plotting correlation heat maps. Principal component analysis (PCA) with scores test is performed which applies a univariate statistical test to the scores of each of the retained principal components to produce a summary statistic for the degree of separation for each pair of classes (Davidson et al. 2016). After data analysis, we receive outlier metabolites but not all of them are annotated properly. MSI levels help us tag meaningful biological information to these metabolites and in communicating the confidence of identifications (Salek et al. 2013) and there are five levels under which these classifications are made (Schrimpe-Rutledge et al. 2016; Salek et al. 2013). Overall, the basic workflow for metabolomics includes preprocessing of data, data normalization, statistical analysis, and finally annotation through various MSI levels (Table 5.1). This is followed by pathway analysis and manual checking of biological relevance for disease-related medicines in significant metabolite list or any other medication depending on the cohort selected. The target metabolites or indirectly affected biomolecules can be validated using the targeted studies. The overall process does face several problems, which includes experimental planning, sample management, and technical changes, among other things, that must be addressed in order to achieve reliable biological data (Box 5.1).

TABLE 5.1
List of Packages for Metabolite Data Analysis and Databases

S. No.	Functionality	Packages
1	Mass spectrometry data handling and (pre-) processing	xcms, MSnbase, ExpressionSet
2	Metabolite identification with MS data	RMassBank, metaMS
3	Statistical analysis of metabolomics data	muma, MOFA, MetaboAnalystR
4	Handling of molecular and chemical structure databases	rcdk, webchem, ClassyFire
5	Network analysis and biochemical pathways	MetaboDiff, pwOmics, MetCirc
6	Multifunctional workflows	SimExTargId, CAMERA, MAIT
7	User interfaces and data sharing	Galaxy, MetaboLights
8	Code-based repositories	CRAN, Bio-C, Git-Hub, rforge.net, sourceforge.net
9	Metabolite raw data accessibility	MetaboLights. Metabolomics Workbench, OmicsDi

BOX 5.1 CHALLENGES AND ADVANCEMENTS IN THE FIELD OF METABOLOMICS AND ITS ROLE IN DIAGNOSTICS PURPOSES (LONG ET AL. 2020; AGGARWAL ET AL. 2021)

Metabolomics Current Challenges	Metabolomics Current Solutions
i. Sample confounding factors: cohort size increases analytical variability increasing biological variability	i. Limit them based on pilot study results, sample injection randomization, regular QC, blank, and IS evaluation
ii. Sample information mismanagement	ii. Sample barcoding and tracking
iii. Sample storage condition variability	iii. Reproducible sampling and storing conditions
iv. Sample degradation and quality deterioration	iv. Immediate quenching and snap freezing, reducing freeze-thaw cycles
v. Sample preparation variations	v. Automated preparations, personnel training, and efficiency
vi. Equipment and unknown failure	vi. Routine maintenance
vii. Instrumental variation	vii. Internal standard (IS), blank and pooled sample assessment
viii. Analytical errors and inter-batch variability	viii. Automation, labeled IS, lesser batch in preparation
ix. Metabolite coverage reduction	ix. Simple, minimal, and fit-for-purpose sample preparation protocol

x. Different separation and ionization techniques and their different strengths and weaknesses	x. The choice should be made accordingly and appropriately based on the sample type and target strategy

COVID-19 Metabolomics Study Progress

- Biospecimens-based metabolome profiling for understanding the systemic effect of COVID-19
- Affected pathways
- Diagnostic and prognostic prediction

COVID-19 Metabolomics Study Prospect

- Unified protocols
- Greater research on comorbidities and asymptomatic cases
- Universal data sharing and meta-analysis with other omics data
- Accessibility of unified data to researchers, clinicians, and industrialists for quick translation of the findings

5.3 Role of metabolome profiling in understanding host response to COVID-19

SARS-CoV-2 in COVID-19 attacks the respiratory tract of the host. Alveolar cells being the first cells to come in contact with the virus (Harrison, Lin, and Wang 2020) result in alarming the host by activation of the innate response, complement cascade by classical, alternative, and lectin pathways (Perico et al. 2021; Battagello et al. 2020). The latter also induces the activation of the coagulation pathway (Margină et al. 2020). The by-products of the activated cascades engender the release of inflammatory biomolecules resulting in a cytokine storm (Shen et al. 2020; Thomas et al. 2020). Cytokine storm results in pulmonary inflammation and activates the coagulation system resulting in excessive clotting and damage to the alveoli barrier (Mahmudpour et al. 2020). The cytokine storm leads to activation of B and T cells that results in further inflammatory biomolecules leading to multi-organ failure (Barberis et al. 2020). The metabolomics studies on minimum to non-invasive biospecimens have been conducted to find the potential metabolite-based diagnostics/prognostics marker candidates or to understand the significantly altered pathways for targeted therapeutics.

Metabolome profiling of plasma and serum-based biospecimens highlighted alterations in kynurenine (Danlos et al. 2021), fatty acid, amino acid (Shen et al. 2020; Song et al. 2020; Thomas et al. 2020), nitrogen (Shen et al. 2020; Thomas et al. 2020), pyrimidine, TCA cycle, fructose and mannose metabolism, and guanosine monophosphate (GMP) of nucleotide biosynthesis pathways, carbamoyl phosphate of the urea cycle, and carbon metabolism (Wu et al. 2020; Y. Zhang et al. 2021).

Alteration in carbon metabolism mainly impacts the TCA cycle, gluconeogenesis, and glycolysis in the severe patient cohort ("CT-Based Staging and Prognosis of Novel Coronavirus (COVID-19) Pneumonia: Correlation with Blood Glucose Levels", 2020). The majorly altered metabolites in the TCA cycle were found to be fumarate, succinate, porphyrin due to the alteration of glycine, and α-ketoglutarate due to alteration of glutamine, histidine, and proline (Barberis et al. 2020). It is also reported that the sulfur-containing amino acid levels were reported to be low while precursor molecules for amino acids such as creatinine, polyamines spermidine, and acetyl-spermidine levels were enhanced in a cohort of COVID-19 patients with medium/high IL-6 level (Shen et al. 2020; Thomas et al. 2020; Lodge et al. 2021). Nan et al. reported the correlation between cytokines and amino acid alteration in rhesus macaques indicating a potential role in therapeutics by regulating cytokine storm in COVID-19 patients (Xiao et al. 2021). Altered phospholipids and amino acid pathways have exhibited post-COVID-19 recovered pulmonary dysfunction as compared to the healthy donor, reported by Juanjuan et al. (Xu et al. 2021). Benjamin et al. reported the role of lipid dysregulation in the immune system moderation in severe COVID-19 patients (Schwarz et al. 2021).

Furthermore, the alteration in carbon molecules resulted in alteration in amino acids, which caused alteration in nitrogen metabolism (Thomas et al. 2020). The study also reported the enhanced expression of α-ketoglutarate in COVID-19 patients suspecting the main reason for the level elevation to be the alteration in transamination activity to homeostasis nitrogen levels in patients (Aggarwal et al. 2021). Minhas et al. and Cameron et al. have reported NAD+ modulates activation of macrophages regulating inflammatory molecules (Minhas et al. 2019; Cameron et al. 2019). Collin et al. (Heer et al. 2020) successfully validated NAD metabolism alteration in COVID-19 patients with the help of cell lines of the human lung (Heer et al. 2020). Ayres et al. reported the development of the diabetic condition in the post-COVID-19 recovered patients due to islet cell damage in the pancreas because of cytokine storm, electrolyte disturbance, muscle hypoxia, and rhabdomyolysis due to metabolite disturbance (Ayres 2020) (Figure 5.3a). Hypoxic condition correlates with amino acid catabolism and alters the saturation levels that may result in lung damage (Páez-Franco et al. 2021). Recovered patients also reported having damaged liver resulting in dyslipidemia (Bai et al. 2021). Metabolome profiling of various samples has led to an understanding of the pathology of the SARS-CoV-2 better (Figure 5.3b).

The diseases from which an individual suffers during his lifetime are the result of the combined effects of the human genome (G) and exposome (E; representing all exposures) (the term "exposome" encompasses all exogenous and endogenous exposures) (Rappaport Stephen M. et al. 2014). But the attribution of risks to G and E and their interaction (G × E) has been problematic because of disparities in characterizing genes and exposures (Rappaport and Smith 2010; Wild 2005). One way to level the playing field would be to explore the health impacts of E and G × E with exposome-wide association studies (EWAS) (Rappaport 2012) that obtain comprehensive, quantitative measurements of chemicals in human biospecimens (Holmes et al. 2008; Ritchie et al. 2010; Wang et al. 2011). This approach recognizes that meaningful exposures are mediated in the internal chemical environment (Rappaport and

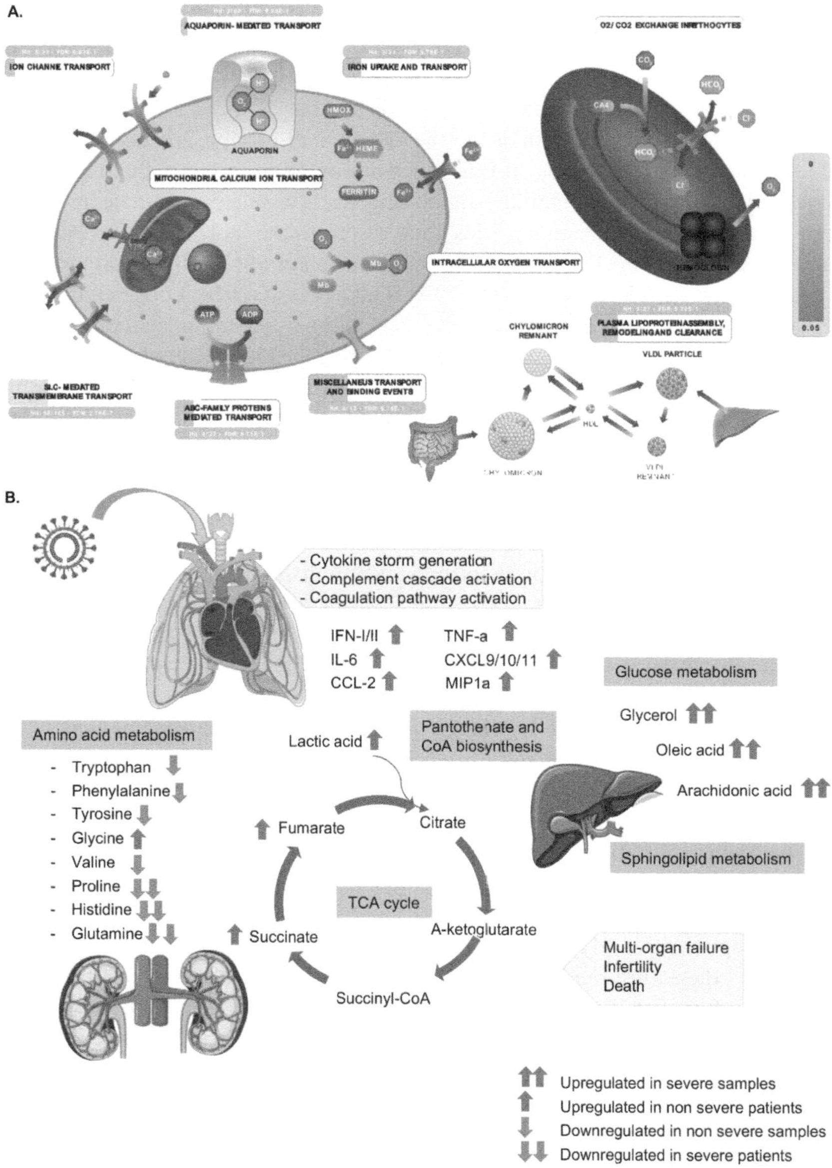

FIGURE 5.3 Effect of COVID-19 in host metabolome profiling. (a) Small molecule transport pathway alteration in COVID-19 patients. (b) Altered pathways in severe and non-severe COVID-19 patients; orange highlights affected pathways and light orange highlights the symptoms or outcome of the altered metabolites and pathways.

Source: This figure was created using Servier Medical Art templates, which are licensed under a Creative Commons Attribution 3.0 Unreported License; https://smart.servier.com and Reactome.

Smith 2010) by endogenous signaling molecules, exogenous chemicals, and reactive electrophiles (E-factors) that communicate with cells, tissues, and organs via mutations, posttranslational modifications, enzymes, transcription factors, and receptors (G-factors) (Ie and R 2009; Liebler 2008; Menon and Manning 2013). Because blood transports chemicals to and from tissues and represents a reservoir of all endogenous and exogenous chemicals in the body at a given time (Nicholson et al. 2012), the blood exposome offers a parsimonious but essentially unexplored means for interrogating biologically relevant exposures (Rappaport 2012). The exposome-wide association study (ExWAS) provides a new approach for conceptualizing the roles and relationships of multiple chemical and non-chemical exposures in the etiology and progression of human diseases at key developmental periods, over the life course (Juarez and Matthews-Juarez 2018). Therefore, the challenge is to characterize these environmental chemicals for their occurrence, fate and transport, effects, assessment, mitigation, and remediation strategies after confident identification, quantification, and association with human health status or disease (Misra 2019).

Single-molecule assays are being replaced with multi-targets assays or untargeted chemical profiling offered by mass spectrometry–based targeted and untargeted metabolomics assays (Marksteiner et al. 2018; Cohen et al. 2018). The increased breadth of the analysis has the potential to provide quantitative data for up to 900 compounds (Barupal et al. 2019; Hu et al. 2019; T. Long et al. 2017; Price et al. 2017), as exemplified in Karl et al. (2017), where untargeted metabolomics was used to compare 737 chemicals in the plasma of soldiers before and after rigorous training exercises. Untargeted plasma metabolomics has been routinely used in clinical and epidemiological settings to identify exposure biomarkers (Rothwell et al. 2014) and chronic disease risk factors (Li et al. n.d.). The extreme complexity and dynamic range of the blood exposome should motivate data-driven studies to discover unknown causes of chronic diseases, regardless of their exogenous and endogenous origins (Rappaport 2012). Candidate exposures can be identified by EWAS that compares omics profiles in blood from diseased and healthy subjects. Thus, there is a need to promote a more global approach to systems biology (Nicholson and Wilson 2003) that expands beyond the endogenous metabolome to the blood exposome, consisting of a large sample of circulating small molecules and inorganic species. A key factor in determining variations in disease severity may be the exposome or measure of all exposures across an individual's lifetime (environmental toxins, pharmacological treatments, lifestyle, diet, etc.) (Naughton et al. 2020). It will be interesting to see how the exposome concept can be added to the metabolomics studies of COVID-19.

5.4 Conclusions and future perspectives

SARS-CoV-2 infection leads to a life-threatening situation in a population with comorbidity but has a long-lasting post-COVID-19 effect on the population who have recovered. SARS-CoV-2 is found to generate cytokine storm and activates immune cascades affecting multiple organs. It also alters the level of amino acids, TCA cycle intermediates, and dysregulates glucose metabolism and lipid metabolism leading to multiple organ damage in COVID-19 patients, which may synchronize with the comorbid condition and infertility in males (Ayres 2020). Metabolite-based diagnostics and prognostic assays may facilitate efficient and quick screening of patient condition for effective treatment. However, exposome analysis of a patient is also an important part to understand the vulnerability or risk factors involved at individual level, against a particular disease. The blood exposome helps to discover all chemicals that cause disease and then to intervene to modify exposures and the concomitant burden of disease (Christiani 2011). Therefore, the application of a comprehensive framework, for metabolome and exposome, may lead to better tailoring and evaluating the effectiveness of public health measures.

REFERENCES

Agache, Ioana, Rachel Miller, James E. Gern, Peter W. Hellings, Marek Jutel, Antonella Muraro, Wanda Phipatanakul, Santiago Quirce, and David Peden. 2019. "Emerging Concepts and Challenges in Implementing the Exposome Paradigm in Allergic Diseases and Asthma: A Practall Document." *Allergy* 74 (3): 449–63. https://doi.org/10.1111/all.13690.

Aggarwal, Shalini, Arup Acharjee, Amrita Mukherjee, Mark S. Baker, and Sanjeeva Srivastava. 2021. "Role of Multiomics Data to Understand Host—Pathogen Interactions in COVID-19 Pathogenesis." *Journal of Proteome Research*, January. https://doi.org/10.1021/acs.jproteome.0c00771.

Álvarez-Sánchez, B., F. Priego-Capote, and M.D. Luque de Castro. 2010. "Metabolomics Analysis II. Preparation of Biological Samples Prior to Detection." *TrAC Trends in Analytical Chemistry* 29 (2): 120–27. https://doi.org/10.1016/j.trac.2009.12.004.

Amir Hashim, Nurul Azmir, Sharaniza Ab-Rahim, Wan Zurinah Wan Ngah, Sheila Nathan, Nurul Syakima Ab Mutalib, Ismail Sagap, A. Rahman A. Jamal, and Musalmah Mazlan. 2020. "Global Metabolomics Profiling of Colorectal Cancer in Malaysian Patients." *BioImpacts* 11 (1): 33–43. https://doi.org/10.34172/bi.2021.05.

Auerswald, Heidi, Sokhoun Yann, Sokha Dul, Saraden In, Philippe Dussart, Nicholas J. Martin, Erik A. Karlsson, and Jose A. Garcia-Rivera. 2021. "Assessment of Inactivation Procedures for SARS-CoV-2." *Journal of General Virology*, January. https://doi.org/10.1099/jgv.0.001539.

Ayres, Janelle S. 2020. "A Metabolic Handbook for the COVID-19 Pandemic." *Nature Metabolism* 2 (7): 572–85. https://doi.org/10.1038/s42255-020-0237-2.

Bai, Yunpeng, Wendong Huang, Yaocai Li, Changchun Lai, Sumei Huang, Guangwen Wang, Yuemei He, Linhui Hu, and Chunbo Chen. 2021. "Lipidomic Alteration of Plasma in Cured COVID-19 Patients Using Ultra High-Performance Liquid Chromatography with High-Resolution Mass Spectrometry." *Bioscience Reports* 41 (BSR20204305). https://doi.org/10.1042/BSR20204305.

Barberis, Elettra, Sara Timo, Elia Amede, Virginia V. Vanella, Chiara Puricelli, Giuseppe Cappellano, Davide Raineri, et al. 2020. "Large-Scale Plasma Analysis Revealed New Mechani=sms and Molecules Associated with the Host Response to SARS-CoV-2." *International Journal of Molecular Sciences* 21 (22). https://doi.org/10.3390/ijms21228623.

Barupal, Dinesh K., Ying Zhang, Tong Shen, Sili Fan, Bryan S. Roberts, Patrick Fitzgerald, Benjamin Wancewicz, et al. 2019. "A Comprehensive Plasma Metabolomics Dataset for a Cohort of Mouse Knockouts within the International Mouse Phenotyping Consortium." *Metabolites* 9 (5): 101. https://doi.org/10.3390/metabo9050101.

Battagello, Daniella S., Guilherme Dragunas, Marianne O. Klein, Ana L. P. Ayub, Fernando J. Velloso, and Ricardo G. Correa. 2020. "Unpuzzling COVID-19: Tissue-Related Signaling Pathways Associated with SARS-CoV-2 Infection and Transmission." *Clinical Science (London, England: 1979)* 134 (16): 2137–60. https://doi.org/10.1042/CS20200904.

Beckonert, Olaf, Hector C. Keun, Timothy M. D. Ebbels, Jacob Bundy, Elaine Holmes, John C. Lindon, and Jeremy K. Nicholson. 2007. "Metabolic Profiling, Metabolomic and Metabonomic Procedures for NMR Spectroscopy of Urine, Plasma, Serum and Tissue Extracts." *Nature Protocols* 2 (11): 2692–703. https://doi.org/10.1038/nprot.2007.376.

Bernatchez, Jean A., and Laura-Isobel McCall. 2020. "Insights Gained into Respiratory Infection Pathogenesis Using Lung Tissue Metabolomics." Edited by Rebecca Ellis Dutch. *PLoS Pathogens* 16 (7): e1008662. https://doi.org/10.1371/journal.ppat.1008662.

Boysen, Angela K., Katherine R. Heal, Laura T. Carlson, and Anitra E. Ingalls. 2018. "Best-Matched Internal Standard Normalization in Liquid Chromatography—Mass Spectrometry Metabolomics Applied to Environmental Samples." *Analytical Chemistry* 90 (2): 1363–69. https://doi.org/10.1021/acs.analchem.7b04400.

Caesar, Lindsay K., Olav M. Kvalheim, and Nadja B. Cech. 2018. "Hierarchical Cluster Analysis of Technical Replicates to Identify Interferents in Untargeted Mass Spectrometry Metabolomics." *Analytica Chimica Acta* 1021 (August): 69–77. https://doi.org/10.1016/j.aca.2018.03.013.

Cameron, Alanna M., Angela Castoldi, David E. Sanin, Lea J. Flachsmann, Cameron S. Field, Daniel J. Puleston, Ryan L. Kyle, et al. 2019. "Inflammatory Macrophage Dependence on NAD + Salvage Is a Consequence of Reactive Oxygen Species—Mediated DNA Damage." *Nature Immunology* 20 (4): 420–32. https://doi.org/10.1038/s41590-019-0336-y.

Cerrato, Andrea, Giuseppe Cannazza, Anna Laura Capriotti, Cinzia Citti, Giorgia La Barbera, Aldo Laganà, Carmela Maria Montone, Susy Piovesana, and Chiara Cavaliere. 2020. "A New Software-Assisted Analytical Workflow Based on High-Resolution Mass Spectrometry for the Systematic Study of Phenolic Compounds in Complex Matrices." *Talanta* 209 (March): 120573. https://doi.org/10.1016/j.talanta.2019.120573.

Chetwynd, Andrew J., Warwick B. Dunn, and Giovanny Rodriguez-Blanco. 2017. "Collection and Preparation of Clinical Samples for Metabolomics." In *Metabolomics: From Fundamentals to Clinical Applications*, edited by Alessandra Sussulini, 965:19–44. Advances in Experimental Medicine and Biology. Cham: Springer International Publishing. https://doi.org/10.1007/978-3-319-47656-8_2.

Christiani, David C. 2011. "Combating Environmental Causes of Cancer." *New England Journal of Medicine* 364 (9): 791–93. https://doi.org/10.1056/NEJMp1006634.

Cohen, Isaac V., Elizabeth T. Cirulli, Matthew W. Mitchell, Thomas J. Jonsson, James Yu, Naisha Shah, Tim D. Spector, Lining Guo, J. Craig Venter, and Amalio Telenti. 2018. "Acetaminophen (Paracetamol) Use Modifies the Sulfation of Sex Hormones." *EBioMedicine* 28 (February): 316–23. https://doi.org/10.1016/j.ebiom.2018.01.033.

Colvin, Heather N., and Regina Joice Cordy. 2020. "Insights into Malaria Pathogenesis Gained from Host Metabolomics." *PLoS Pathogens* 16 (11). https://doi.org/10.1371/journal.ppat.1008930.

Costa dos Santos Junior, Gilson, Claudia Maria Pereira, Tatiana Kelly da Silva Fidalgo, and Ana Paula Valente. 2020. "Saliva NMR-Based Metabolomics in the War Against COVID-19." *Analytical Chemistry* 92 (24): 15688–92. https://doi.org/10.1021/acs.analchem.0c04679.

"CT-Based Staging and Prognosis of Novel Coronavirus (COVID-19) Pneumonia: Correlation with Blood Glucose Levels." 2020. *European Review (blog)*. December 22, 2020. www.europeanreview.org/article/24212.

Danlos, François-Xavier, Claudia Grajeda-Iglesias, Sylvère Durand, Allan Sauvat, Mathilde Roumier, Delphine Cantin, Emeline Colomba, et al. 2021. "Metabolomic Analyses of COVID-19 Patients Unravel Stage-Dependent and Prognostic Biomarkers." *Cell Death & Disease* 12 (3): 1–11. https://doi.org/10.1038/s41419-021-03540-y.

Darnell, Miriam E. R., Kanta Subbarao, Stephen M. Feinstone, and Deborah R. Taylor. 2004. "Inactivation of the Coronavirus That Induces Severe Acute Respiratory Syndrome, SARS-CoV." *Journal of Virological Methods* 121 (1): 85–91. https://doi.org/10.1016/j.jviromet.2004.06.006.

Davidson, R. L., R. J. Weber, H. Liu, A. Sharma-Oates, and M. R. Viant. 2016. "Galaxy-M: A Galaxy Workflow for Processing and Analyzing Direct Infusion and Liquid Chromatography Mass Spectrometry-Based Metabolomics Data." *Gigascience* 5 (10). https://doi.org/10.1186/s13742-016-0115-8.

Doğan, Halef O., Onur Şenol, Serkan Bolat, Şeyma N. Yıldız, Seyit A. Büyüktuna, Rağıp Sarıismailoğlu, Kübra Doğan, Mürşit Hasbek, and Süleyman N. Hekim. 2020. "Understanding the Pathophysiological Changes via Untargeted Metabolomics in COVID-19 Patients." *Journal of Medical Virology*, December. https://doi.org/10.1002/jmv.26716.

Godzien, Joanna, Vanesa Alonso-Herranz, Coral Barbas, and Emily Grace Armitage. 2015. "Controlling the Quality of Metabolomics Data: New Strategies to Get the Best Out of the QC Sample." *Metabolomics* 11 (3): 518–28. https://doi.org/10.1007/s11306-014-0712-4.

Gong, Zhi-Gang, Jing Hu, Xi Wu, and Yong-Jiang Xu. 2017. "The Recent Developments in Sample Preparation for Mass Spectrometry-Based Metabolomics." *Critical Reviews in Analytical Chemistry* 47 (4): 325–31. https://doi.org/10.1080/10408347.2017.1289836.

Gorrochategui, Eva, Joaquim Jaumot, Sílvia Lacorte, and Romà Tauler. 2016. "Data Analysis Strategies for Targeted and Untargeted LC-MS Metabolomic Studies: Overview and Workflow." *TrAC Trends in Analytical Chemistry* 82 (September): 425–42. https://doi.org/10.1016/j.trac.2016.07.004.

Gough, Derrick V., H. Daniel Bahaghighat, and Robert E. Synovec. 2019. "Column Selection Approach to Achieve a High Peak Capacity in Comprehensive Three-Dimensional Gas Chromatography." *Talanta* 195 (April): 822–29. https://doi.org/10.1016/j.talanta.2018.12.007.

Harrison, Andrew G., Tao Lin, and Penghua Wang. 2020. "Mechanisms of SARS-CoV-2 Transmission and Pathogenesis." *Trends in Immunology* 41 (12): 1100–15. https://doi.org/10.1016/j.it.2020.10.004.

Heer, Collin D., Daniel J. Sanderson, Lynden S. Voth, Yousef M. O. Alhammad, Mark S. Schmidt, Samuel A. J. Trammell, Stanley Perlman, Michael S. Cohen, Anthony R. Fehr, and Charles Brenner. 2020. "Coronavirus Infection and PARP Expression Dysregulate the NAD Metabolome: An Actionable Component of Innate Immunity." *Journal of Biological Chemistry*, October. https://doi.org/10.1074/jbc.RA120.015138.

Hemati, Maral, Mohsen Soosanabadi, Tahereh Ghorashi, Hadi Ghaffari, Azadeh Vahedi, Elaheh Sabbaghian, Zahra Rasouli Nejad, Amir Salati, Navid Danaei, and Parviz Kokhaei. 2020. "Thermal Inactivation of COVID-19 Specimens Improves RNA Quality and Quantity." *Journal of Cellular Physiology*, December. https://doi.org/10.1002/jcp.30206.

Hessling, Martin, Katharina Hoenes, and Christian Lingenfelder. 2020. "Selection of Parameters for Thermal Coronavirus Inactivation—a Data-Based Recommendation." *GMS Hygiene and Infection Control* 15: Doc16, July. https://doi.org/10.3205/DGKH000351.

Heßling, Martin, Katharina Hönes, Petra Vatter, and Christian Lingenfelder. 2020. "Ultraviolet Irradiation Doses for Coronavirus Inactivation—Review and Analysis of Coronavirus Photoinactivation Studies." *GMS Hygiene and Infection Control* 15: Doc08, May. https://doi.org/10.3205/DGKH000343.

Hewavitharana, Amitha K. 2011. "Matrix Matching in Liquid Chromatography—Mass Spectrometry with Stable Isotope Labelled Internal Standards—Is It Necessary?" *Journal of Chromatography A* 1218 (2): 359–61. https://doi.org/10.1016/j.chroma.2010.11.047.

Holmes, Elaine, Ruey Leng Loo, Jeremiah Stamler, Magda Bictash, Ivan K. S. Yap, Queenie Chan, Tim Ebbels, et al. 2008. "Human Metabolic Phenotype Diversity and Its Association with Diet and Blood Pressure." *Nature* 453 (7193): 396–400. https://doi.org/10.1038/nature06882.

Hu, Jiun-Ruey, Morgan E. Grams, Josef Coresh, Seungyoung Hwang, Csaba P. Kovesdy, Eliseo Guallar, Eugene P. Rhee, and Tariq Shafi. 2019. "Serum Metabolites and Cardiac Death in Patients on Hemodialysis." *Clinical Journal of the American Society of Nephrology: CJASN* 14 (5): 747–49. https://doi.org/10.2215/CJN.12691018.

Ie, Brodsky, and R. Medzhitov. 2009. "Targeting of Immune Signalling Networks by Bacterial Pathogens." *Nature Cell Biology* 11 (5): 521–26. https://doi.org/10.1038/ncb0509-521.

Instituto Tecnológico de Aguascalientes, and Miguel Sánchez-Brito. 2020. "Saliva Analysis Using FTIR Spectroscopy to Detect Possible SARS-CoV-2 (COVID-19) Virus Carriers." *Revista Mexicana de Ingeniería Biomédica*. https://doi.org/10.17488/RMIB.42.1.1.

Jain, Abhishek, Xin Hui Li, and Wei Ning Chen. 2019. "An Untargeted Fecal and Urine Metabolomics Analysis of the Interplay between the Gut Microbiome, Diet and Human Metabolism in Indian and Chinese Adults." *Scientific Reports* 9 (1): 9191. https://doi.org/10.1038/s41598-019-45640-y.

Jang, Sukbin, Si Hyun Han, and Ji-Young Rhee. 2020. "Cluster of Coronavirus Disease Associated with Fitness Dance Classes, South Korea." *Emerging Infectious Diseases* 26 (8): 1917–20. https://doi.org/10.3201/eid2608.200633.

Joice Cordy, Regina. 2020. "Mining the Human Host Metabolome Toward an Improved Understanding of Malaria Transmission." *Frontiers in Microbiology* 11 (February). https://doi.org/10.3389/fmicb.2020.00164.

Juarez, Paul D., and Patricia Matthews-Juarez. 2018. "Applying an Exposome-Wide (ExWAS) Approach to Cancer Research." *Frontiers in Oncology* 8. https://doi.org/10.3389/fonc.2018.00313.

Karl, J. Philip, Lee M. Margolis, Nancy E. Murphy, Christopher T. Carrigan, John W. Castellani, Elisabeth H. Madslien, Hilde-Kristin Teien, Svein Martini, Scott J. Montain, and Stefan M. Pasiakos. 2017. "Military Training Elicits Marked Increases in Plasma Metabolomic Signatures of Energy Metabolism, Lipolysis, Fatty Acid Oxidation, and Ketogenesis." *Physiological Reports* 5 (17): e13407. https://doi.org/10.14814/phy2.13407.

Lakshmanan, Viswanathan, Kyu Y. Rhee, and Johanna P. Daily. 2011. "Metabolomics and Malaria Biology." *Molecular and Biochemical Parasitology* 175 (2): 104–11. https://doi.org/10.1016/j.molbiopara.2010.09.008.

Li, Xinmin S., Zeneng Wang, Tomas Cajka, Jennifer A. Buffa, Ina Nemet, Alex G. Hurd, Xiaodong Gu, et al. n.d. "Untargeted Metabolomics Identifies Trimethyllysine, a TMAO-Producing Nutrient Precursor, as a Predictor of Incident Cardiovascular Disease Risk." *JCI Insight* 3 (6). Accessed March 26, 2018. https://doi.org/10.1172/jci.insight.99096.

Liebler, Daniel C. 2008. "Protein Damage by Reactive Electrophiles: Targets and Consequences." *Chemical Research in Toxicology* 21 (1): 117–28. https://doi.org/10.1021/tx700235t.

Locher, Kerstin, Billie Velapatino, Mélissa Caza, Lisa Li, Charlene Porter, and Marthe Charles. 2020. "Approach to Assessment of New Swabs and Viral Transport Media for SARS-CoV-2 Testing." *Journal of Clinical Microbiology*, November. https://doi.org/10.1128/JCM.01562-20.

Lodge, Samantha, Philipp Nitschke, Torben Kimhofer, Jerome D. Coudert, Sofina Begum, Sze-How Bong, Toby Richards, et al. 2021. "NMR Spectroscopic Windows on the Systemic Effects of SARS-CoV-2 Infection on Plasma Lipoproteins and Metabolites in Relation to Circulating Cytokines." *Journal of Proteome Research* 20 (2): 1382–96. https://doi.org/10.1021/acs.jproteome.0c00876.

Long, Nguyen Phuoc, Tran Diem Nghi, Yun Pyo Kang, Nguyen Hoang Anh, Hyung Min Kim, Sang Ki Park, and Sung Won Kwon. 2020. "Toward a Standardized Strategy of Clinical Metabolomics for the Advancement of Precision Medicine." *Metabolites* 10 (2): 51. https://doi.org/10.3390/metabo10020051.

Long, Tao, Michael Hicks, Hung-Chun Yu, William H. Biggs, Ewen F. Kirkness, Cristina Menni, Jonas Zierer, et al. 2017. "Whole-Genome Sequencing Identifies Common-to-Rare Variants Associated with Human Blood Metabolites." *Nature Genetics* 49 (4): 568–78. https://doi.org/10.1038/ng.3809.

Luies, Laneke, Ilse du Preez, and Du Toit Loots. 2017. "The Role of Metabolomics in Tuberculosis Treatment Research." *Biomarkers in Medicine* 11 (11): 1017–29. https://doi.org/10.2217/bmm-2017-0141.

Mahmud, Iqbal, and Timothy J. Garrett. 2020. "Mass Spectrometry Techniques in Emerging Pathogens Studies: COVID-19 Perspectives." *Journal of the American Society for Mass Spectrometry* 31 (10): 2013–24. https://doi.org/10.1021/jasms.0c00238.

Mahmudpour, Mehdi, Jamshid Roozbeh, Mohsen Keshavarz, Shokrollah Farrokhi, and Iraj Nabipour. 2020. "COVID-19 Cytokine Storm: The Anger of Inflammation." *Cytokine* 133 (September): 155151. https://doi.org/10.1016/j.cyto.2020.155151.

Maras, Jaswinder Singh, Shvetank Sharma, Adil Bhat, Reshu Aggrawal, Ekta Gupta, and Shiv K. Sarin. 2020. "Multi-Omics Integration Analysis of Respiratory Specimen Characterizes Baseline Molecular Determinants Associated with COVID-19 Diagnosis." Preprint. *Infectious Diseases (Except HIV/AIDS)*. https://doi.org/10.1101/2020.07.06.20147082.

Margină, Denisa, Anca Ungurianu, Carmen Purdel, George Mihai Nițulescu, Dimitris Tsoukalas, Evangelia Sarandi, Maria Thanasoula, et al. 2020. "Analysis of the Intricate Effects of Polyunsaturated Fatty Acids and Polyphenols on Inflammatory Pathways in Health and Disease." *Food and Chemical Toxicology: An International Journal Published for the British Industrial Biological Research Association* 143 (September): 111558. https://doi.org/10.1016/j.fct.2020.111558.

Marksteiner, Josef, Imrich Blasko, Georg Kemmler, Therese Koal, and Christian Humpel. 2018. "Bile Acid Quantification of 20 Plasma Metabolites Identifies Lithocholic Acid as a Putative Biomarker in Alzheimer's Disease." *Metabolomics* 14 (1). https://doi.org/10.1007/s11306-017-1297-5.

Martins, Duarte L., Jure Sencar, Nikolaus Hammerschmidt, Andreas Flicker, Johanna Kindermann, Thomas R. Kreil, and Alois Jungbauer. 2020. "Truly Continuous Low PH Viral Inactivation for Biopharmaceutical Process Integration." *Biotechnology and Bioengineering* 117 (5): 1406–17. https://doi.org/10.1002/bit.27292.

Menon, Suchithra, and Brendan D. Manning. 2013. "Cell Signalling: Nutrient Sensing Lost in Cancer." *Nature* 498 (June): 444–45. https://doi.org/10.1038/498444a.

Miller, Ian J., Sean R. Peters, Katherine A. Overmyer, Brett R. Paulson, Michael S. Westphall, and Joshua J. Coon. 2019. "Real-Time Health Monitoring through Urine Metabolomics." *NPJ Digital Medicine* 2 (1): 109. https://doi.org/10.1038/s41746-019-0185-y.

Minhas, Paras S., Ling Liu, Peter K. Moon, Amit U. Joshi, Christopher Dove, Siddhita Mhatre, Kevin Contrepois, et al. 2019. "Macrophage de Novo NAD+ Synthesis Specifies Immune Function in Aging and Inflammation." *Nature Immunology* 20 (1): 50–63. https://doi.org/10.1038/s41590-018-0255-3.

Misra, Biswapriya B. 2019. "Metabolomics Tools to Study Links Between Pollution and Human Health: An Exposomics Perspective." *Current Pollution Reports* 5 (3): 93–111. https://doi.org/10.1007/s40726-019-00109-4.

Naughton, Sean X., Urdhva Raval, Joyce M. Harary, and Giulio M. Pasinetti. 2020. "The Role of the Exposome in Promoting Resilience or Susceptibility after SARS-CoV-2 Infection." *Journal of Exposure Science & Environmental Epidemiology* 30 (5): 776–77. https://doi.org/10.1038/s41370-020-0232-4.

Nicholson, Jeremy K., Elaine Holmes, James M. Kinross, Ara W. Darzi, Zoltan Takats, and John C. Lindon. 2012. "Metabolic Phenotyping in Clinical and Surgical Environments." *Nature* 491 (7424): 384–92. https://doi.org/10.1038/nature11708.

Nicholson, Jeremy K., and Ian D. Wilson. 2003. "Understanding 'Global' Systems Biology: Metabonomics and the Continuum of Metabolism." *Nature Reviews Drug Discovery* 2 (8): 668–76. https://doi.org/10.1038/nrd1157.

Oliver, Stephen G., Michael K. Winson, Douglas B. Kell, and Frank Baganz. 1998. "Systematic Functional Analysis of the Yeast Genome." *Trends in Biotechnology* 16 (9): 373–78. https://doi.org/10.1016/S0167-7799(98)01214-1.

Páez-Franco, José C., Jiram Torres-Ruiz, Víctor A. Sosa-Hernández, Rodrigo Cervantes-Díaz, Sandra Romero-Ramírez, Alfredo Pérez-Fragoso, David E. Meza-Sánchez, et al. 2021. "Metabolomics Analysis Reveals a Modified Amino Acid Metabolism That Correlates with Altered Oxygen Homeostasis in COVID-19 Patients." *Scientific Reports* 11 (1): 6350. https://doi.org/10.1038/s41598-021-85788-0.

Pastorino, Boris, Franck Touret, Magali Gilles, Xavier de Lamballerie, and Remi N. Charrel. 2020. "Heat Inactivation of Different Types of SARS-CoV-2 Samples: What Protocols for Biosafety, Molecular Detection and Serological Diagnostics?" *Viruses* 12 (7): 735. https://doi.org/10.3390/v12070735.

Patterson, Edward I., Tessa Prince, Enyia R. Anderson, Aitor Casas-Sanchez, Shirley L. Smith, Cintia Cansado-Utrilla, Tom Solomon, et al. 2020. "Methods of Inactivation of SARS-CoV-2 for Downstream Biological Assays." *The Journal of Infectious Diseases* 222 (9): 1462–67. https://doi.org/10.1093/infdis/jiaa507.

Perico, Luca, Ariela Benigni, Federica Casiraghi, Lisa F. P. Ng, Laurent Renia, and Giuseppe Remuzzi. 2021. "Immunity, Endothelial Injury and Complement-Induced Coagulopathy in COVID-19." *Nature Reviews. Nephrology* 17 (1): 46–64. https://doi.org/10.1038/s41581-020-00357-4.

Preez, Ilse du, Laneke Luies, and Du Toit Loots. 2017. "Metabolomics Biomarkers for Tuberculosis Diagnostics: Current Status and Future Objectives." *Biomarkers in Medicine* 11 (2): 179–94. https://doi.org/10.2217/bmm-2016-0287.

Price, Nathan D., Andrew T. Magis, John C. Earls, Gustavo Glusman, Roie Levy, Christopher Lausted, Daniel T. McDonald, et al. 2017. "A Wellness Study of 108 Individuals Using Personal, Dense, Dynamic Data Clouds." *Nature Biotechnology* 35 (8): 747–56. https://doi.org/10.1038/nbt.3870.

Pristner, Manuel, and Benedikt Warth. 2020. "Drug—Exposome Interactions: The Next Frontier in Precision Medicine." *Trends in Pharmacological Sciences* 41 (12): 994–1005. https://doi.org/10.1016/j.tips.2020.09.012.

Rampler, Evelyn, Harald Schoeny, Bernd M. Mitic, Yasin El Abiead, Michaela Schwaiger, and Gunda Koellensperger. 2018. "Simultaneous Non-Polar and Polar Lipid Analysis by on-Line Combination of HILIC, RP and High Resolution MS." *The Analyst* 143 (5): 1250–58. https://doi.org/10.1039/C7AN01984J.

Rappaport, Stephen M. 2012. "Biomarkers Intersect with the Exposome." *Biomarkers : Biochemical Indicators of Exposure, Response, and Susceptibility to Chemicals* 17 (6): 483–89. https://doi.org/10.3109/1354750X.2012.691553.

Rappaport, Stephen M., Dinesh K. Barupal, David Wishart, Paolo Vineis, and Augustin Scalbert. 2014. "The Blood Exposome and Its Role in Discovering Causes of Disease." *Environmental Health Perspectives* 122 (8): 769–74. https://doi.org/10.1289/ehp.1308015.

Rappaport, Stephen M., and Martyn T. Smith. 2010. "Environment and Disease Risks." *Science (New York, N.Y.)* 330 (6003): 460–61. https://doi.org/10.1126/science.1192603.

Richardson, Safiya, Jamie S. Hirsch, Mangala Narasimhan, James M. Crawford, Thomas McGinn, Karina W. Davidson, and the Northwell COVID-19 Research Consortium et al. 2020. "Presenting Characteristics, Comorbidities, and Outcomes Among 5700 Patients Hospitalized with COVID-19 in the New York City Area." *JAMA* 323 (20): 2052. https://doi.org/10.1001/jama.2020.6775.

Ritchie, Shawn A., Pearson W. K. Ahiahonu, Dushmanthi Jayasinghe, Doug Heath, Jun Liu, Yingshen Lu, Wei Jin, et al. 2010. "Reduced Levels of Hydroxylated, Polyunsaturated Ultra Long-Chain Fatty Acids in the Serum of Colorectal Cancer Patients: Implications for Early Screening and Detection." *BMC Medicine* 8 (1): 13. https://doi.org/10.1186/1741-7015-8-13.

Rodino, Kyle G., Mark J. Espy, Seanne P. Buckwalter, Robert C. Walchak, Jeffery J. Germer, Emily Fernholz, Aimee Boerger, Audrey N. Schuetz, Joseph D. Yao, and Matthew J. Binnicker. 2020. "Evaluation of Saline, Phosphate-Buffered Saline, and Minimum Essential Medium as Potential Alternatives to Viral Transport Media for SARS-CoV-2 Testing." Edited by Alexander J. McAdam. *Journal of Clinical Microbiology* 58 (6): e00590–20. https://doi.org/10.1128/JCM.00590-20.

Roessner, Ute, and Jairus Bowne. 2009. "What Is Metabolomics All About?" *BioTechniques* 46 (5): 363–65. https://doi.org/10.2144/000113133.

Rogers, Amy A., Russell E. Baumann, Gwynngelle A. Borillo, Ron M. Kagan, Hollis J. Batterman, Marzena M. Galdzicka, and Elizabeth M. Marlowe. 2020. "Evaluation of Transport Media and Specimen Transport Conditions for the Detection of SARS-CoV-2 by Use of Real-Time Reverse Transcription-PCR." Edited by Alexander J. McAdam. *Journal of Clinical Microbiology* 58 (8): e00708–20. https://doi.org/10.1128/JCM.00708-20.

Rothwell, Joseph A., Yoann Fillâtre, Jean-François Martin, Bernard Lyan, Estelle Pujos-Guillot, Leopold Fezeu, Serge Hercberg, et al. 2014. "New Biomarkers of Coffee Consumption Identified by the Non-Targeted Metabolomic Profiling of Cohort Study Subjects." *PLoS ONE* 9 (4): e93474. https://doi.org/10.1371/journal.pone.0093474.

Salek, Reza M., Christoph Steinbeck, Mark R. Viant, Royston Goodacre, and Warwick B. Dunn. 2013. "The Role of Reporting Standards for Metabolite Annotation and Identification in Metabolomic Studies." *GigaScience* 2 (1): 13–13. https://doi.org/10.1186/2047-217X-2-13.

Salinas, Jorge L., Jessica C. Kissinger, Dean P. Jones, and Mary R. Galinski. 2014. "Metabolomics in the Fight against Malaria." *Memórias Do Instituto Oswaldo Cruz* 109 (5): 589–97. https://doi.org/10.1590/0074-0276140043.

Sauerschnig, Claudia, Maria Doppler, Christoph Bueschl, and Rainer Schuhmacher. 2017. "Methanol Generates Numerous Artifacts during Sample Extraction and Storage of Extracts in Metabolomics Research." *Metabolites* 8 (1): 1. https://doi.org/10.3390/metabo8010001.

Schrimpe-Rutledge, Alexandra C., Simona G. Codreanu, Stacy D. Sherrod, and John A. McLean. 2016. "Untargeted Metabolomics Strategies—Challenges and Emerging Directions." *Journal of the American Society for Mass Spectrometry* 27 (12): 1897–905. https://doi.org/10.1021/jasms.8b05178.

Schwaiger, Michaela, Harald Schoeny, Yasin El Abiead, Gerrit Hermann, Evelyn Rampler, and Gunda Koellensperger. 2019. "Merging Metabolomics and Lipidomics into One Analytical Run." *The Analyst* 144 (1): 220–29. https://doi.org/10.1039/C8AN01219A.

Schwarz, Benjamin, Lokesh Sharma, Lydia Roberts, Xiaohua Peng, Santos Bermejo, Ian Leighton, Arnau Casanovas-Massana, et al. 2021. "Cutting Edge: Severe SARS-CoV-2 Infection in Humans Is Defined by a Shift in the Serum Lipidome, Resulting in Dysregulation of Eicosanoid Immune Mediators." *The Journal of Immunology* 206 (2): 329–34. https://doi.org/10.4049/jimmunol.2001025.

Segers, Karen, Sven Declerck, Debby Mangelings, Yvan Vander Heyden, and Ann Van Eeckhaut. 2019. "Analytical Techniques for Metabolomic Studies: A Review." *Bioanalysis* 11 (24): 2297–318. https://doi.org/10.4155/bio-2019-0014.

Shen, Bo, Xiao Yi, Yaoting Sun, Xiaojie Bi, Juping Du, Chao Zhang, Sheng Quan, Fangfei Zhang, Rui Sun, Liujia Qian, et al. 2020a. "Proteomic and Metabolomic Characterization of COVID-19 Patient Sera." *Cell* 182 (1): 59–72.e15. https://doi.org/10.1016/j.cell.2020.05.032.

Shi, Ding, Ren Yan, Longxian Lv, Huiyong Jiang, Yingfeng Lu, Jifang Sheng, Jiaojiao Xie, et al. 2021. "The Serum Metabolome of COVID-19 Patients Is Distinctive and Predictive." *Metabolism* 118 (May): 154739. https://doi.org/10.1016/j.metabol.2021.154739.

Snytnikova, Olga A., Anastasiya A. Khlichkina, Renad Z. Sagdeev, and Yuri P. Tsentalovich. 2019. "Evaluation of Sample Preparation Protocols for Quantitative NMR-Based Metabolomics." *Metabolomics* 15 (6): 84. https://doi.org/10.1007/s11306-019-1545-y.

Song, Jin-Wen, Sin Man Lam, Xing Fan, Wen-Jing Cao, Si-Yu Wang, He Tian, Gek Huey Chua, et al. 2020. "Omics-Driven Systems Interrogation of Metabolic Dysregulation in COVID-19 Pathogenesis." *Cell Metabolism* 32 (2): 188–202.e5. https://doi.org/10.1016/j.cmet.2020.06.016.

Stokvis, Ellen, Hilde Rosing, and Jos H. Beijnen. 2005. "Stable Isotopically Labeled Internal Standards in Quantitative Bioanalysis Using Liquid Chromatography/Mass Spectrometry: Necessity or Not?" *Rapid Communications in Mass Spectrometry* 19 (3): 401–07. https://doi.org/10.1002/rcm.1790.

Thomas, Tiffany, Davide Stefanoni, Julie A. Reisz, Travis Nemkov, Lorenzo Bertolone, Richard O. Francis, Krystalyn E. Hudson, et al. 2020. "COVID-19 Infection Alters Kynurenine and Fatty Acid Metabolism, Correlating with IL-6 Levels and Renal Status." *American Society for Clinical Investigation*, July 23, 2020. https://doi.org/10.1172/jci.insight.140327.

Trivedi, Drupad K., Katherine A. Hollywood, and Royston Goodacre. 2017. "Metabolomics for the Masses: The Future of Metabolomics in a Personalized World." *New Horizons in Translational Medicine* 3 (6): 294–305. https://doi.org/10.1016/j.nhtm.2017.06.001.

Tsutsui, Haruhito, Toshio Maeda, Jun Zhe Min, Shinsuke Inagaki, Tatsuya Higashi, Yoshiyuki Kagawa, and Toshimasa Toyo'oka. 2011. "Biomarker Discovery in Biological Specimens (Plasma, Hair, Liver and Kidney) of Diabetic Mice Based upon Metabolite Profiling Using Ultra-Performance Liquid Chromatography with Electrospray Ionization Time-of-Flight Mass Spectrometry." *Clinica Chimica Acta; International Journal of Clinical Chemistry* 412 (11–12): 861–72. https://doi.org/10.1016/j.cca.2010.12.023.

Wang, Zeneng, Elizabeth Klipfell, Brian J. Bennett, Robert Koeth, Bruce S. Levison, Brandon DuGar, Ariel E. Feldstein, et al. 2011. "Gut Flora Metabolism of Phosphatidylcholine Promotes Cardiovascular Disease." *Nature* 472 (7341): 57–63. https://doi.org/10.1038/nature09922.

Welch, Stephen R., Katherine A. Davies, Hubert Buczkowski, Nipunadi Hettiarachchi, Nicole Green, Ulrike Arnold, Matthew Jones, et al. 2020. "Analysis of Inactivation of SARS-CoV-2 by Specimen Transport Media, Nucleic Acid Extraction Reagents, Detergents, and Fixatives." Edited by Michael J. Loeffelholz. *Journal of Clinical Microbiology* 58 (11): e01713–20. https://doi.org/10.1128/JCM.01713-20.

Wild, Christopher Paul. 2005. "Complementing the Genome with an 'Exposome': The Outstanding Challenge of Environmental Exposure Measurement in Molecular Epidemiology." *Cancer Epidemiology and Prevention Biomarkers* 14 (8): 1847–50. https://doi.org/10.1158/1055-9965.EPI-05-0456.

Wild, Christopher Paul. 2012. "The Exposome: From Concept to Utility." *International Journal of Epidemiology* 41 (1): 24–32. https://doi.org/10.1093/ije/dyr236.

Wu, Di, Ting Shu, Xiaobo Yang, Jian-Xin Song, Mingliang Zhang, Chengye Yao, Wen Liu, et al. 2020. "Plasma Metabolomic and Lipidomic Alterations Associated with COVID-19." *National Science Review* 7 (7): 1157–68. https://doi.org/10.1093/nsr/nwaa086.

Xiao, Nan, Meng Nie, Huanhuan Pang, Bohong Wang, Jieli Hu, Xiangjun Meng, Ke Li, et al. 2021. "Integrated Cytokine and Metabolite Analysis Reveals Immunometabolic Reprogramming in COVID-19 Patients with Therapeutic Implications." *Nature Communications* 12 (1): 1618. https://doi.org/10.1038/s41467-021-21907-9.

Xu, Juanjuan, Mei Zhou, Ping Luo, Zhengrong Yin, Sufei Wang, Tingting Liao, Fan Yang, et al. 2021. "Plasma Metabolomic Profiling of Patients Recovered From Coronavirus Disease 2019 (COVID-19) With Pulmonary Sequelae 3 Months After Discharge." *Clinical Infectious Diseases*, no. ciab147 (February). https://doi.org/10.1093/cid/ciab147.

Yuan, Min, Susanne B. Breitkopf, Xuemei Yang, and John M. Asara. 2012. "A Positive/Negative Ion—Switching, Targeted Mass Spectrometry—Based Metabolomics Platform for Bodily Fluids, Cells, and Fresh and Fixed Tissue." *Nature Protocols* 7 (5): 872–81. https://doi.org/10.1038/nprot.2012.024.

Zeki, Özge Cansın, Cemil Can Eylem, Tuba Reçber, Sedef Kır, and Emirhan Nemutlu. 2020. "Integration of GC—MS and LC—MS for Untargeted Metabolomics Profiling." *Journal of Pharmaceutical and Biomedical Analysis* 190 (October): 113509. https://doi.org/10.1016/j.jpba.2020.113509.

Zhang, Aihua, Hui Sun, Guangli Yan, Ping Wang, and Xijun Wang. 2015. "Metabolomics for Biomarker Discovery: Moving to the Clinic." *BioMed Research International* 2015. https://doi.org/10.1155/2015/354671.

Zhang, Yuchen, Rui Guo, Sharon H. Kim, Hardik Shah, Shuting Zhang, Jin Hua Liang, Ying Fang, et al. 2021. "SARS-CoV-2 Hijacks Folate and One-Carbon Metabolism for Viral Replication." *Nature Communications* 12 (1): 1676. https://doi.org/10.1038/s41467-021-21903-z.

Zheng, Fujian, Xinjie Zhao, Zhongda Zeng, Lichao Wang, Wangjie Lv, Qingqing Wang, and Guowang Xu. 2020. "Development of a Plasma Pseudotargeted Metabolomics Method Based on Ultra-High-Performance Liquid Chromatography—Mass Spectrometry." *Nature Protocols* 15 (8): 2519–37. https://doi.org/10.1038/s41596-020-0341-5.

6

Protein microarrays for COVID-19 research

Biomarker discovery, humoral response, and vaccine targets

Arup Acharjee, Abhilash Barpanda, Jing Ren, and Xiaobo Yu

CONTENTS

DOI: 10.1201/9781003220787-6

List of abbreviations

COVID-19 – Coronavirus disease 2019
ELISA – Enzyme-linked immunosorbent assay
LC-MS/MS – liquid chromatography with tandem mass spectrometry
MS – Mass spectrometry
PhIP-seq – Phage immunoprecipitation sequencing
SARS-CoV-2 – Severe acute respiratory syndrome coronavirus 2

6.1 Introduction

6.1.1 Introduction to the microarray-based techniques

Studying protein and proteomics has been a discipline of interest among researchers from different fields. Protein microarray, unlike other system biology techniques, is very simple and less time-consuming technique. The principle of protein arrays itself makes the technique very much dynamic and flexible, which can be made available for clinical research in an economical way.

6.1.2 Principle of microarray-based research and diagnostics

Protein microarray-based technologies have been used widely for over a decade to investigate the differential genetic and proteomic expression in pathogens. This technique has been applied widely to detect pathogens, pathogen discovery, microbial resistance, and strain typing (Natesan and Ulrich 2010). Besides, it is widely used for examining host responses to disease and discovering therapeutics. Proteome microarray is a rapidly growing area capable of being a robust diagnostic platform for innumerable diseases (S. Li et al. 2021).

A protein microarray approach for diagnosing infectious diseases such as SARS-CoV-2 could be an excellent way of detecting the infection as these arrays are highly sensitive and accurate. The SARS-CoV-2 virus has several known antigenic proteins such as the spike glycoprotein and nucleocapsid phosphoprotein present on its outer surface. Specific analytical arrays with surface-specific antibodies can easily detect the presence of the virus in samples in less than an hour. The diagnostic workflow is as simple as an ELISA assay and requires a significantly less volume of samples, and can perform multiple samples at once.

6.1.3 Advantages of microarray over MS technique for clinical diagnostics

Of all the technological interventions used to probe the COVID-19 biological sample, namely, serum or plasma, microarrays have provided unique and unbiased data about the biology of SARS-CoV-2 infection in greater detail (Aggarwal et al. 2021). Protein microarrays provide an attractive format for the detection and diagnosis of wide range of analytes. It provides the following advantages: (a) simplified workflow, (b) high-throughput platform, (c) straightforward translation of proteomics results to clinics, and (d) easy scale-up.

6.1.4 Use of microarrays in other disease conditions

Technologies such as proteome arrays (Davies et al. 2005) have been instrumental in assessing host response in multiple viral diseases such as papillomavirus (Luevano et al. 2010), immunization against rubella virus (Haralambieva et al. 2017), the discovery of immune markers, and vaccine targets against herpes simplex virus 1 and 2 (Kalantari-Dehaghi et al. 2012) and the immune response against the Zika virus (Mishra, Uversky, and Giri 2018). Hence it was natural to use these platforms to understand COVID-19 biology.

6.2 Types of protein microarrays

6.2.1 Protein microarray technology

Three basic types of protein microarrays—analytical microarray, functional microarray, and reverse-phase microarray—are predominantly used to investigate proteins' biochemical properties (Figure 6.1).

6.2.2 Analytical protein microarray

Analytical protein microarray is a powerful technique having a great potential to aid in the detection and identification of a wide range of analytes in various applications such as clinical diagnostics, proteomics, drug development, and molecular cell biology. Antibody arrays are the most representative form of an analytical array (Sutandy et al. 2013). The analytical microarray can be used to study various biological processes such as protein–protein interactions, signal transduction cascades, posttranslational modification of proteins, detection of toxins such as snake venoms (Sauer 2017). These arrays have also revealed several advantages as compared to classical single-stage Western blotting or ELISA assays in clinical settings as arrays are high-throughput, extremely sensitive, inexpensive, and require a very small volume of samples for multiple protein detection (Figure 6.2).

6.2.3 Functional protein microarray

The functional protein microarrays are made of full-length functional proteins or functional protein domains. Fabricating a functional array is quite delicate. Unlike DNA, *in vitro* protein synthesis is not very robust. Besides, protein folding needs to be accurate to be functional. Simultaneously, fabricating all these factors needs to be considered carefully for optimization of the immobilization of proteins to the slide surface. Because proteins printed are intact, various biochemical properties of the proteins, such as interactions with biomolecules, chemical-binding activities,

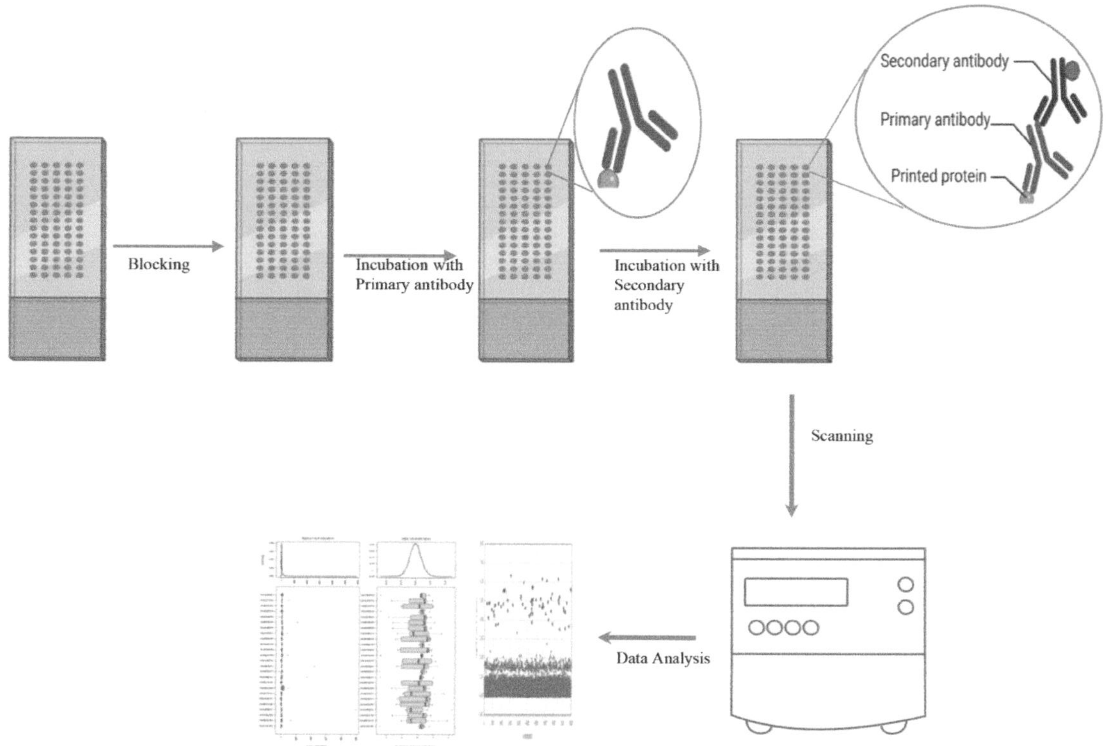

FIGURE 6.1 Principle of protein microarray. Protein microarray is a powerful high-throughput technique having a great potential to aid in the detection and identification of a wide range of analytes in various applications, including clinical research and diagnosis. The microarray analytes detection procedure starts with blocking the microarray slides to inactivate the empty surfaces on the glass slide. Once the blocking process completes, the next important step is the addition of biological samples containing primary antibodies. This step allows binding of the antibodies present in the sample with the specific protein printed on the chip. To detect the primary antibody, a secondary antibody tagged with fluorophores can be added, followed by scanning and detection of the potential markers. Mostly the microarray chip contains a huge number of printed proteins that generate big data upon experimentation. The data needs to be checked carefully using biostatistics and bioinformatics tools to interpret the data, and hence it is the ultimate step in such kind of analysis.

and enzyme kinetics, can be investigated using these arrays (Romanov et al. 2014). Functional arrays are used extensively to investigate protein–protein, protein–peptide, protein–DNA, protein–RNA, protein–microRNA, protein–small molecule, etc. interactions (Hu et al. 2011). The role of functional arrays has become indispensable in studying posttranslational modifications of proteins such as phosphorylation, glycosylation, ubiquitination, and acetylation.

BOX 6.1 PROTEIN MICROARRAY FOR COVID-19 RESEARCH: FOCUS ON BIOMARKER DISCOVERY AND DIAGNOSTICS

A protein microarray approach for the diagnosis of infectious diseases such as SARS-CoV-2 would be an excellent way of detecting infection as these arrays are highly sensitive and accurate. The SARS-CoV-2 virus has several known antigenic proteins such as the spike glycoprotein and nucleocapsid phosphoprotein present on its outer surface, which can be useful for COVID-19 diagnosis. The diagnostic workflow is as simple as an ELISA assay and requires a very less volume of samples, and it allows performing multiple samples at once. The methodology follows with the spotting of GST tagged monoclonal antibodies on the activated surface followed by confirmation of the capturing process. GST tags present on the antibody can be used to check the quality of the slides. Once the slide passes quality control checks, it can be used for the detection of viral proteins in the given samples, such as plasma, serum, saliva swab, etc. Blocking the slides with proteins such as BSA is the next step. Once blocking is done, slides are ready for sample overlay followed by a brief incubation and then the addition of the same monoclonal antibody tagged with biotin to make a sandwich pattern. Streptavidin-based detection reagents are added after that to detect the presence of viral protein. Using the strategy, several studies were performed for the identification of diagnostic markers and potential candidates for vaccine development. A study involving the detection of unique IgG antibodies generated against 13 recombinant proteins linked to 4 structural proteins, including S, E, N, and M proteins of SARS-CoV, suggested that anti-N and anti-S antibodies can be used as a diagnostic marker of COVID-19 infection. It is also identified that a component of protein S is immunogenic, and it could be a potential candidate for the development of vaccines. Proteins from the SARS coronavirus (SARS-CoV) and five other coronaviruses were included in a protein microarray created by Zhu et al. They also created a computer algorithm that predicts samples from SARS patients using multiple classifiers, which they used to predict 206 sera from Chinese fever patients. Patients with antibodies to other coronavirus proteins were also detected by the examination. It can be concluded from these studies that protein microarrays can be used to detect viral-specific marker antibodies in sera on a large scale in a responsive, fast, and simple manner and can be used for diagnostic tests.

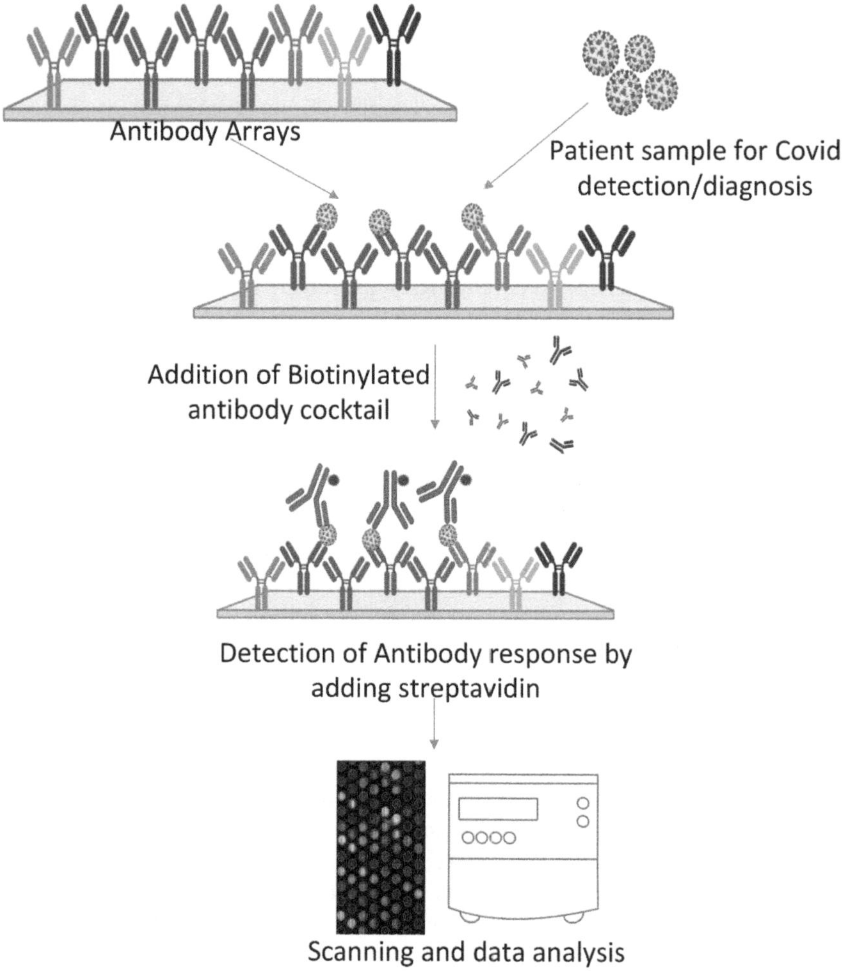

FIGURE 6.2 Principle of antibody microarray. Antibody arrays are a distinct semiquantitative form of protein microarray. The array measures multiple proteins saving time, cost, and sample consumption. The array with antibody against proteins of SARS-Cov-2 can be used to detect the abundance of multiple analytes with reference to COVID-19 detection in biological samples, using different label-based assays and sandwich assays. In a typical antibody microarray experiment, the most crucial step is the immobilization of the analytes, i.e., desired antibody onto a chemically modified surface. The first step of the microarray experimentation is the blocking of the reactive groups of the surface. After the blocking step, COVID-19 suspect samples such as a nasopharyngeal swab, plasma, or serum that can be added and incubated for binding and captured by the antibody. The subsequent binding can be screened either by fluorescent labeling of the samples or by the addition of suitable reagents.

6.2.4 Reverse-phase protein microarray

The reverse-phase expands microarray technology by employing a very generalized format to the primary platform. The term "reverse-phase" denotes that the antigen is used as a capture entity instead of an antibody. It can be used to investigate biological processes in diverse animal models starting from drosophila to humans with various biological inputs such as cell lines, tissue samples, serum, plasma, and tissue microdissection (Boellner and Becker 2015). It is exclusively suitable for sketching the *in vivo* signal transduction cascades (Uzoma and Zhu 2013) as it requires minimal total cell extracts, extraordinary sensitivity (picomole to femtomole range), and exceptional accuracy. Quantitative investigation of phosphorylated, glycosylated, acetylated, or whole cellular proteins from multiple samples can be achieved with the reverse-phase protein microarrays (Zhao et al. 2009).

6.2.5 Peptide microarrays

Compared to DNA and protein microarrays, peptide arrays are more challenging but most instructive for the studies mentioned previously as they enable us to narrow down from the massive protein stretch to an exact binding domain (Katz et al. 2011). Peptide arrays have multiple advantages in studying proteomic research as peptides are easy to be synthesized when the protein sequence is available. In a peptide array, probes or linker groups such as His-tag, and GST, biotin can be quickly introduced at any given position and can be used for various detection assays (Figure 6.3). Because folding is not an issue here, quality control of these arrays becomes more advantageous. At the same time, a peptide does not truly replicate the entire polypeptide structure; hence, peptide arrays should be used where it fits exclusively. A chemically manufactured peptide microarray representing the principal antigens of the virus can be built on a glass slide for the detection of antibody responses (Hansen, Buus, and Schafer-Nielsen 2013). The assay would predict antigenic peptides with high sensitivity and specificity to identify potential epitopes for vaccine development. Spotting purified protein domains can be an alternative for making a peptide microarray. Arrays printed with domains would also be potent in recognition of potential epitopes for designing a vaccine.

6.2.6 VirScan

The concept of VirScan technique was introduced by Xu et al. VirScan is based on the phage immunoprecipitation sequencing (PhIP-seq) technology, and the method uses a bacteriophage library that exhibits proteome-wide peptides FROM a wide range of pathogenic viruses. The VirScan library comprises approximately 115,760 distinct peptide sets representing known pathogenic human viruses and several non-viral antigens collected from Immune Epitope Database (Hasan et al. 2021).

To investigate the serological profiling using VirScan, serum samples are overlayed with the library. Mixing of serum samples with the VirScan library allows antibodies from the serum to bind with specific pathogenic epitopes present on the bacteriophage surface of the VirScan. The phage–antibody complexes are further immunoprecipitated, followed by identifying peptides and respective DNA regions encoding the peptides. The DNA is further sequenced by NGS, revealing multiple information about the viral infections (Shrock et al. 2020). Following the sequencing, results are interpreted with statistical analysis of the obtained data to identify potential antibodies against a wide range of pathogens. The data analysis relies on statistical evaluation (p values) of the relative enrichment of pathogen-specific peptides present in the serum samples. Therefore, the results obtained are purely qualitative and are not directly equivalent to other conventional quantitative singleplex serological techniques. VirScan combined with NGS sequencing has emerged as a powerful new technique for high-throughput serological screening, as it can correctly detect commonly encountered antiviral antibodies with higher sensitivity and specificity (95%) compared to regular ELISA and Western blot assays.

6.3 Sample preparation and experimental workflow for COVID-19 studies

6.3.1 Viral inactivation

SARS-CoV-2 being a highly infectious agent, biofluid samples like serum require care during the handling. Viral inactivation is achieved by heat treatment of the samples at 56°C for 30 minutes (Darnell et al. 2004).

6.3.2 Viral proteome array workflow

The construction of the viral proteome array is quite challenging, and several essential steps need to be considered carefully. First of all, to increase the genetic contents, the predicted ORFs are amplified using primers that contain appropriate recombination sites and then cloned into suitable expression systems (Qi et al. 2017). The expressed proteins are then purified and followed by the fabrication of a microarray slide.

BOX 6.2 PROTEIN MICROARRAYS FOR COVID-19 RESEARCH: FOCUS ON EPITOPE MAPPING

Extensive profiling of humoral antibody responses to SARS-Cov-2 proteins is important in the understanding of host immunity. The majority of serological studies to date have identified a general humoral response to SARS-CoV-2 infection against the spike (S) and the nucleocapsid (N) proteins, which can be seen as early as the fourth day after symptom onset when SARS-CoV-2-specific IgG and IgM antibodies appear simultaneously. The peptide epitope mapping of the humoral response would provide a true landscape of antigenicity of SARS-CoV-2 proteins and could be useful in diagnostics and vaccine development. Cataloging of viral peptides is crucial for the making of a peptide array for epitope mapping. For the detection of antibody responses, a chemically manufactured peptide representing the virus's core antigens can be installed on a glass slide followed by overlaying the host convalescent sera for the binding of potential antibodies to the printed peptides. Following a similar strategy, Musicò et al. (2021) have identified the immunodeterminant regions present on SARS-CoV-2 spike protein, Orf1ab polyprotein, and nucleocapsid proteins. Using the viral microarray, Wang et al. (2020) have identified the landscape of B-cell epitopes against SARS-CoV-2-specific antibodies in host sera.

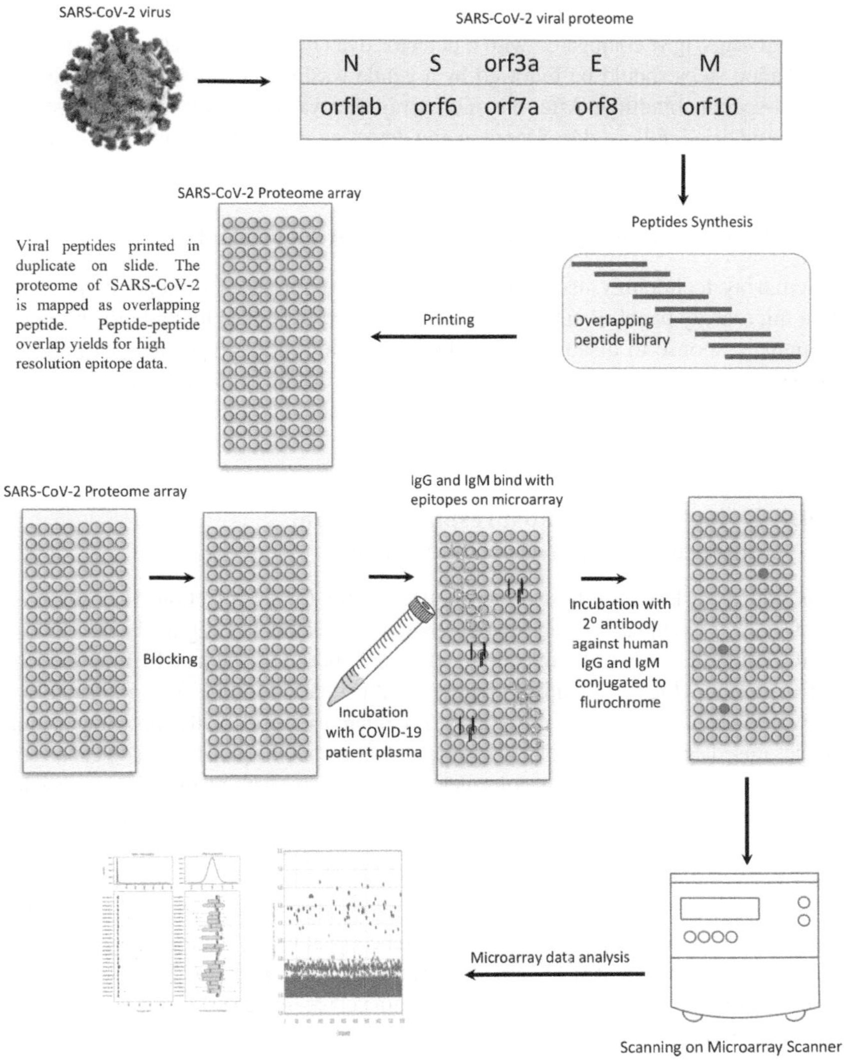

FIGURE 6.3 SARS-CoV-2 microarray. The viral proteome can be mimicked with the construction of a proteome-wide peptide array. The peptide array can be used for identification of serological markers as well as epitope mapping. Incubation of COVID-19 patient plasma samples with the array followed by detection with secondary antibody cocktail can give wide information about the humoral immune response against the virus. A similar approach can be used to identify the potential epitope peptide for vaccine design.

As mentioned earlier, once fabrication is done, the experiment is seemingly straightforward. The microarray chip is initially blocked using BSA blocking solution on a gentle shake. In this step, the non-specific protein BSA is adsorbed to the microarray slides' vacant surfaces and helps improve the signal-to-noise ratio. After blocking, the array is incubated with a serum sample followed by detecting the antibodies reacting with the spotted proteins on the microarray using antihuman IgG/IgA/IgM conjugated with a detector dye (Jiang et al. 2020). Each of these incubation steps should be followed by a gentle washing step, reducing the chance of non-specific bindings. After the microarray was washed, it was rinsed in distilled water and dried, followed by scanning and detecting the altered antibody responses.

6.4 Protein microarray applications in COVID-19

Microarray technology uses antibodies against hundreds of proteins or peptides in a microarray panel or pathogen proteome, where it is used to study the humoral immune response to discovering antibodies of multiple isotypes against multiple antigens (Heiss et al. 2020). Quite naturally, for COVID-19 studies, researchers have used slides with both printed antibodies and overlapping peptides of the SARS-CoV-2 proteome.

6.4.1 Detection of the host response to viral infection using protein microarray

It is critical to learn the host response after SARS-CoV-2 infection. Host response knowledge is essential to establish a diagnostic and therapeutic regime backed by scientific evidence to clear the infection and reduce mortality adequately. In a study on patient serum, Hou et al. studied serum of 132 patients suffering from influenza and COVID-19 by using a protein microarray. The microarray essentially consisting of antibodies raised against a panel of 532 serological proteins. The panel's serological targets included intercellular signaling molecules, protein-binding activity modulators, protein-modifying enzymes, and metabolite interconversion enzymes. The in-depth profiling of the patient serum could identify an extensive set of differentially expressed proteins that showed around it perturbed 26 biological processes during the early stage of COVID-19. In brief, the most upregulated signaling pathways during early COVID-19 included cytokine–cytokine receptor interaction, cytokine signaling in the immune system, IL-4 and IL-13 signaling, and IL17 signaling. The JAK-STAT pathway, MAP kinase, and PI3K-Akt signaling pathways were also upregulated during acute SARS-CoV-2 infection. However, few pathways like neutrophil degranulation, complement cascade, and coagulation cascades were downregulated during the initial stage of COVID-19 disease. Primarily, these pathways indicated the dysregulation of inflammation pathways and immune signaling during SARS-CoV-2 infection. Furthermore, a significant correlation was established between neutrophils with CCL2 and lymphocytes with CXCL10 mediated cytokine signaling pathways (Hou et al. 2020).

6.4.2 Detection of humoral response using microarray for immunogenicity

Another host response that is essential for tracking disease pathogenesis is the immune response. Microarrays provide a matchless platform to study the immunogenic humoral response to the pathogen proteome. Thus, it holds the key to understand the epitope diversity during the humoral response after infection at the amino acid level. Wang et al. studied the IgG- and IgM-based B cell response on the SARS-CoV-2 protein. Using the ORF sequences translated from the SARS-CoV-2 genome, a SARS-CoV-2 proteome microarray was developed that contains 966 peptides representing the SARS-CoV-2 proteome and purified nucleoprotein (N), spike (S), and E proteins (Wang et al. 2020). On the array, each peptide was 15 amino acids long, having a 5 amino acid overlap, thus enabling the panoramic detection of thousands of anti-SARS-CoV-2 antibodies in the serum of COVID-19 patients within 1.5 hours at amino acid resolution. Using serum from ten patients, they found 61 IgG and IgM antibody epitopes distributed in the seven SARS-CoV-2 proteins (M, N, S, Orf1ab, Orf3a, Orf7a, and Orf8.) Of all the proteins, it was observed that Orf1ab was found to have the most immunogenic epitopes constituting a total of 32 IgM and IgG epitopes. The Orf1ab proteins which constituted the immunogenic proteins were nsp1–4, nsp6, nsp8–10, and nsp12–16. Additionally, binding epitopes were also identified on eight spikes (S), eight on nucleoprotein (N), ($n = 8$), five on membrane (M) ($n = 5$), four on Orf3a ($n = 4$), three on Orf7a ($n = 3$), and one on Orf8 ($n = 1$) proteins. Notably, four immune-dominant epitopes with antibodies in more than 80% of the COVID-19 patients were present in the N (residue 206–210, SPARM), S (residue 816–820, SFIED), and Orf3a (residue 136–140, KNPLL; residue 176–180, SPISE) proteins. Surprisingly, antibodies to E, Orf6, and Orf10 were not detected. Subsequently, they analyzed the longitudinal changes of humoral immunity in 49 critical COVID-19 patients (Cheng et al. 2021). This work found that the SARS-CoV-2 antibodies produced in patients were related to the severity of COVID-19 infection. Furthermore, the immunogenic epitopes related to the prognosis of COVID-19 were identified, in which the antibodies targeting nsp3 and nsp5 proteases were associated with survival rates, while IgG antibodies targeting structural proteins (N, S, ORF3a) were associated with the risk of mortality in COVID-19 patients.

Another exciting study (Shrock et al. 2020) showed the use of a high-throughput method, VirScan, to study epitopes of antiviral antibodies generated in human sera. The idea was to check for IgG and IgA immunogenicity of SARS-CoV-2 and cross-reactivity between SARS-CoV-2 and other endemic coronaviruses. Also, the study provided a detailed overview of previous viral infections on immune response repertoire. Using COVID19 serum ($n = 232$) and pre-COVID-19 serum (190) on VirScan supplemented with coronavirus libraries, they assayed about 108 antibody repertoire–peptide interactions. The investigators could map 823 epitopes across the entire SARS-CoV-2 proteome that were immunogenic. Ten epitopes could produce neutralizing antibodies, thus good candidates for vaccine candidates or therapies such as a monoclonal antibody therapy for passive vaccination. Also, they report a cross-reactive epitope in the S2 region that is potentially neutralizing and might putatively determine severity.

Another study showed the efficacy of proteome arrays (de Assis et al. 2021) in the detection of immunogenic responses. Using a multiplex antigen array with antigens from SARS-CoV-2, SARS-CoV, MERS-CoV, common endemic human coronavirus strains, and other common respiratory viruses, they studied the IgG and IgM kinetics of COVID-19 patients (de Assis et al. 2021). They found IgG antibodies to be more useful as discriminatory than IgM antibodies, although the overall reactivity pattern was identical. The pre-pandemic sera response was compared to convalescent sera of COVID-19 patients. A clear demarcation was observed in immunogenicity, especially by the spike and nucleoprotein of SARS-COV-2; hence, it can act as excellent biomarkers of active case detection, rapid antibody test, and vaccine development. Using convalescent plasma from COVID-19 patient's microarray designed from 200 linear peptides of the SARS-CoV-2 spike ectodomain, Farrera-Soler et al. (2020) found that there were three immunodominant IgG epitopes corresponding to the amino acid positions 655–672, 787–822, and 1147–1158. A couple of these epitopes relate to the crucial proteolytic sites on the spike protein (S1/S2 and S2′) recognized to be essential for cellular entry. Further, they conducted an alanine substitution scan to understand individual amino acid's contribution and verified the amino acids essential for binding with the antibodies common were H655, Y660, and C662. Such studies provide critical information for vaccine design and also provides information about immune escape variants, as we are already witnessing.

A more thorough approach was adopted by a group of Chinese researchers wherein they probed the entire SARS-CoV-2 proteome for immunodominant epitopes, namely, IgG and IgM (Jiang et al. 2020). They fabricated a proteome microarray having 18 SARS-CoV-2 proteins and studied the IgG and IgM responses on sera from a cohort of 29 convalescent patients. They reported that most patients turned seropositive for the viral N protein and S1 protein. Moreover, immunogenicity was observed in ORF9b and NSP5. Using clinical correlates of patients, they demonstrated that the S1-specific IgG immunogenicity clearly correlated with increasing age and circulating lactate dehydrogenase levels and negatively correlates with lymphocyte percentage. Therefore, this exhibited the potential of being used as a classifier of infection severity as well. Interestingly, the two of the control sera exhibited strong IgG bindings, with the N protein, showing that the N protein is a weak candidate for serological testing due to its evolutionarily conserved nature in coronaviruses.

Using a peptide microarray designed against the spike protein region of the virus (Li et al. 2021), they studied the immunoglobulin G (IgG) immune response landscape from the serum response of 1,051 patients. They reported that the C-terminal domain (CTD) and a region surrounding the S20 cleavage site and fusion peptide are enriched with linear epitopes. CTD had six highly immunogenic epitopes S1–93, S1–97, S1–105/106, S1–111, and S1–113. Surprisingly, the receptor-binding domain (RBD) essential for neutralizing antibody generation does not have linear epitope. Outside of the RBD area, a group of 16 highly immunogenic epitopes was discovered. However, the number of responsive peptides significantly varied among subjects and correlated with illness severity. Especially they found that the patients who could not survive COVID-19 displayed a statistically significant decrease in the IgG response for the epitopes corresponding to S1–93, S1–97, and S2–78. Through the SARS-CoV-2 proteome microarray and spike protein–peptide microarray developed in their laboratory, not only can it fully monitor the body's humoral immune response process. It can also be found that the dominant protein antigens and the dominant B cell epitopes located in the S protein region that can induce neutralizing antibodies contribute to the preliminary evaluation of vaccines and provide strong support for vaccine development. Recently, the same group examined the antibody profile of 59 serum samples from 32 patients received with the COVID-19 inactivated vaccine BBIBP-CorV. It was found that the expression of neutralization antibodies correlated well with the microarray-based antibody response. Furthermore, the antibodies to S1–5 and S2–22 peptides may have the potential to evaluate the effectiveness of inactivated virus vaccines, While the antibodies to N protein, NSP7 and S2–78 may have the value to distinguish COVID-19 patients from vaccinated individuals (Ma et al. 2021).

Using whole SARS-CoV-2 proteome peptide microarrays, Dahlke et al. (2020) tested serum samples for SARS-CoV-2-specific IgA, IgG, and IgM antibodies. They noted that shortly after disease onset, although a clear SARS-CoV-2-specific IgA was noted in moderate cases, a case with more severe progression exhibited a delayed but ultimately extremely intense and comprehensive SARS-CoV-2-specific IgA response. On a cohort of PCR-confirmed mild ($n = 9$) and severe ($n = 7$) COVID-19 patients, scientists studied the IgA and IgG responses in the sera (Schwarz et al. 2021). They used a microarray platform PEPperCHIP® by PEPperPRINT GmbH, Germany, having the proteome of Wuhan-Hu-1 SARS-COV-2 isolate. Fifteen amino acid peptides with 13 amino acid overlap of the viral proteome represented by proteins ORF1a/b, spike, ORF3a, envelope, membrane glycoprotein, ORF6, ORF7a, ORF8, nucleocapsid phosphoprotein, and ORF10 were used. While observing the longitudinal epitope-specific antibody responses across the SARS-CoV-2 proteome, they found a short-lived IgA but IgG response was steady.

Early IgA-specific epitope responses were relatively feeble but peaked at week 3 and then steadily declined till diminishing by week 10 after the infection. However, no IgG response was observed till week 1 of infection. It changed after the first week, and IgG response could be observed well into the late convalescent phase, especially for IgG antibodies against NSP13 peptide E5466-E5480. They further studied if epitopes could also delineate the disease severity based on the immune response. It was observed that IgG antibody against peptide P1622-H1636 (NSP3) was most potent to classify severe from mild COVID-19. The NSP12-derived peptide N4542-D4556 followed it. A less potent but significant severe COVID-19 marker was the NSP3-derived IgA epitope (Y1906-Y1920). Further, they identified three IgG and IgA shared epitopes recognized in sera of patients with severe COVID-19. Of these, the ORF3a-derived peptide Y141-D155 was found to be the most suitable biomarker of severity. This information could be essential for the development of rapid antibody tests for disease severity and progression.

In a study in Italy by Musico and group using the microarray platform PEPperCHIP® by PEPperPRINT GmbH, Germany, having the proteome of Wuhan-Hu-1 SARS-COV-2 isolate on seven blood samples of COVID-19 patients found that the N protein epitope (region 155–171) had excellent diagnostic efficiency in distinguishing COVID-19 positive from healthy persons. Using this epitope, IgG identification in COVID-19 specimens achieved 92% sensitivity and 100% specificity (Musicò et al. 2021). Most importantly, they had no cross-reactivity with endemic coronaviruses. Furthermore, working on the IgM response in sample collected within one month after the initial symptoms, they found that IgM reactivity showed discriminative power. Generally, N protein epitopes 155–171 was identified as a promising candidate for any further study and efficient use in immunoassays.

A multicenter study by Mishra et al., scientists used a high-density microarray to study humoral response across coronaviruses (Mishra et al. 2021). They used plasma samples collected at two time points from a cohort of 50 COVID-19 patients and compared them with 11 patients with IgG antibodies to SARS-CoV-1 who were tested in 2004–2005, 11 patients with IgG antibodies to other seasonal human coronaviruses (HCoV), and 10 healthy human volunteers were taken as control. They could identify 29 linear epitopes from the SARS-CoV-2 that showed immunoreactivity, and could detect from individuals who had mild or severe SARS-CoV-2 infections. They were also able to detect asymptomatic infections in this study.

Unfortunately, SARS-CoV-2 is mutating, with new variants emerging nearly every week that could impede drug development and reduce the efficacy of COVID-19 vaccines. Therefore, the ability to comprehensively map the binding epitopes of neutralizing antibodies (NAbs) to spike variants is critical in stopping the spread of COVID-19. To address this challenge, Zhang et al. prepared an SARS-CoV-2 spike protein variants microarray containing 72 spike protein variants, and developed a multiplexed Spike-ACE2 Inhibitor Screening (mSAIS) assay method based on the microarray, high-throughput detection of the neutralization effect of antiviral antibodies in the serum of convalescent COVID-19 patients and vaccinees on different types of spike protein variants. The results indicate that the high titer neutralizing antibodies that inhibit the interactions of ACE2 with spike variants more efficiently displayed with more diverse epitopes on spike proteins. These results suggest mSAIS platform may have great potential in advancing COVID-19 drug and vaccine development, especially as the SARS-CoV-2 virus continues to evolve (Zhang et al. 2021) are shown in Table 6.1.

TABLE 6.1

Microarray in COVID-19 Research

S. No.	Microarray Type	Sample	Parameter Checked	Key Findings	Reference
1	Whole SARS-CoV-2 proteome peptide microarrays	COVID-19 serum	SARS-CoV-2-specific IgA, IgG, and IgM antibodies	SARS-CoV-2-specific IgA was early and low in moderate COVID-19 and delayed and high in severe COVID-19	Dahlke et al. 2020
2	The whole proteome of Wuhan-Hu-1 SARS-COV-2 isolate	COVID-19 serum	SARS-CoV-2-specific IgA and IgG antibodies	Short-lived IgA but IgG response steady IgG antibody against peptide P1622-H1636 (NSP3) was most potent to classify severe from mild COVID-19 Severe COVID-19 marker was the NSP3-derived IgA epitope (Y1906-Y1920)	Schwarz et al. 2021

3	Microarray of antibodies of 532 biomarkers	COVID-19 serum	Serum biomarkers such as intercellular signaling molecules, protein-binding activity modulators, protein-modifying enzymes, and metabolite interconversion enzymes	Dysregulation of inflammation pathways and immune signaling during SARS-CoV-2 infection	Hou et al. 2020
4	Microarray of 966 peptides representing the SARS-CoV-2 proteome	COVID-19 serum	Tested for IgG and IgM antibody response	61 IgG and IgM antibody epitopes distributed in the seven SARS-CoV-2 proteins M, N, S, Orf1ab, Orf3a, Orf7a, and Orf8	Wang et al. 2020
5	VirScan supplemented with SARS-CoV-2 antigen library	COVID-19 serum and pre-COVID-19 serum	IgG and IgA immunogenicity of SARS-CoV-2 and cross-reactivity between SARS-CoV-2 and other endemic coronaviruses	823 immunogenic epitopes and a cross-reactive epitope in the S2 region might putatively determine severity	Shrock et al. 2020
6	Multiplex antigen array having antigens from SARS-CoV-2, as well as SARS-CoV, MERS-CoV, common endemic human coronavirus strains, and other common respiratory viruses	COVID-19 serum	IgG and IgM kinetics of COVID-19 patients	IgG antibodies to be more useful as discriminatory than IgM antibodies. The spike and nucleoprotein of SARS-CoV-2 showed clear discrimination for active case detection	de Assis et al. 2021
7	Microarray of 200 linear peptides of the SARS-CoV-2 spike ectodomain	Convalescent plasma of COVID-19 patients	IgG response	Three immunodominant IgG epitopes having amino acid positions 655–672, 787–822, and 1147–1158. Further, amino acids essential for binding with the antibodies common were H655, Y660, and C662	Farrera-Soler et al. 2020

(Continued)

TABLE 6.1 (Continued)

S. No.	Microarray Type	Sample	Parameter Checked	Key Findings	Reference
8	Entire SARS-CoV-2 proteome	Convalescent plasma of COVID-19 patients	Immunodominant epitopes for IgG and IgM.	S1-specific IgG immunogenicity correlated with disease severity N protein is a weak candidate for serological testing	Jiang et al. 2020
9	Microarray of 211 linear peptides of the SARS-CoV-2 spike protein	COVID-19 serum	IgG response	C-terminal domain (CTD) and a region surrounding the S20 cleavage site and fusion peptide are both dense in linear epitopes CTD had six highly immunogenic epitopes S1–93, S1–97, S1–105/106, S1–111, and S1–113	Li et al. 2021
10	Entire SARS-CoV-2 proteome	Seven COVID-19 patients, aged 68–96 years, at about 30 days after the onset of symptoms		N protein (region 155–171) candidate for diagnostics and could discriminate COVID-19 positive versus healthy individuals	Musico et al. 2021
11	Microarray of 72 SARS-CoV-2 spike protein variants	Convalescent COVID-19 patients and vaccinees		Serum with high neutralizing antibody titer neutralizes the S protein mutant more widely	Zhang et al. 2021

6.5 Conclusions and authors' perspective

Protein microarrays have proven its utility to study clinical samples simply, reliably, and without complex processing steps. The information gained allows incredibly detailed analysis of the host immune response at the peptide level against the novel pathogen like SARS-CoV-2, thereby vindicating its importance in an era where other omics technology such as mass spectrometry has dominated the scene. The ease of use of protein microarrays also allows for the easy scalability and bench to bedside translation of the knowledge gained and paves the way for developing sensitive lateral flow assays and better vaccines. With the emergence of newer variants with a potential immune escape, more studies need to be conducted using microarrays, increasing our knowledge about the pathogen and host response. Microarrays have thus proved their worth in the fight against the pandemic, and these techniques hold immense potential for their usage against future outbreaks. The technology is still in its infancy, and broader adoption by the scientific community against infectious diseases apart from COVID-19 would be interesting. Humanity is under constant threat from emerging infectious diseases, and, therefore, techniques such as protein microarrays would be one of the essential weapons against the emerging diseases.

REFERENCES

Aggarwal, Shalini, Arup Acharjee, Amrita Mukherjee, Mark S. Baker, and Sanjeeva Srivastava. 2021. "Role of Multiomics Data to Understand Host—Pathogen Interactions in COVID-19 Pathogenesis." *Journal of Proteome Research* 20 (2): 1107–32. https://doi.org/10.1021/acs.jproteome.0c00771.

Assis, Rafael R. de, Aarti Jain, Rie Nakajima, Algis Jasinskas, Jiin Felgner, Joshua M. Obiero, Philip J. Norris, et al. 2021. "Analysis of SARS-CoV-2 Antibodies in COVID-19 Convalescent Blood Using a Coronavirus Antigen Microarray." *Nature Communications* 12 (1): 6. https://doi.org/10.1038/s41467-020-20095-2.

Boellner, Stefanie, and Karl-Friedrich Becker. 2015. "Reverse Phase Protein Arrays—Quantitative Assessment of Multiple Biomarkers in Biopsies for Clinical Use." *Microarrays* 4 (2): 98–114. https://doi.org/10.3390/microarrays4020098.

Cheng, Linlin, Xiaomei Zhang, Yu Chen, Dan Wang, Dong Zhang, et al. 2021. "Dynamic Landscape Mapping of Humoral Immunity to SARS-CoV-2 Identifies Non-structural Protein Antibodies Associated with the Survival of Critical COVID-19 Patients." *Signal Transduction and Targeted Therapy* 6 (1): 304. https://doi.org/10.1038/s41392-021-00718-w.

Dahlke, Christine, Jasmin Heidepriem, Robin Kobbe, René Santer, Till Koch, Anahita Fathi, My L. Ly, et al. 2020. "Distinct Early IgA Profile May Determine Severity of COVID-19 Symptoms: An Immunological Case Series." *medRxiv*. https://doi.org/10.1101/2020.04.14.20059733.

Davies, D. Huw, Xiaowu Liang, Jenny E. Hernandez, Arlo Randall, Siddiqua Hirst, Yunxiang Mu, Kimberly M. Romero, et al. 2005. "Profiling the Humoral Immune Response to Infection by Using Proteome Microarrays: High-Throughput Vaccine and Diagnostic Antigen Discovery." *Proceedings of the National Academy of Sciences* 102 (3): 547–52. https://doi.org/10.1073/pnas.0408782102.

Darnell, Miriam E. R., Kanta Subbarao, Stephen M. Feinstone, and Deborah R. Taylor. 2004. "Inactivation of the Coronavirus That Induces Severe Acute Respiratory Syndrome, SARS-CoV." *Journal of Virological Methods* 121 (1): 85–91. https://doi.org/10.1016/j.jviromet.2004.06.006.

Farrera-Soler, Lluc, Jean-Pierre Daguer, Sofia Barluenga, Oscar Vadas, Patrick Cohen, Sabrina Pagano, Sabine Yerly, Laurent Kaiser, Nicolas Vuilleumier, and Nicolas Winssinger. 2020. "Identification of Immunodominant Linear Epitopes from SARS-CoV-2 Patient Plasma." *PLoS ONE* 15 (9): e0238089. https://doi.org/10.1371/journal.pone.0238089.

Hansen, Lajla Bruntse, Soren Buus, and Claus Schafer-Nielsen. 2013. "Identification and Mapping of Linear Antibody Epitopes in Human Serum Albumin Using High-Density Peptide Arrays." Edited by Morten Nielsen. *PLoS ONE* 8 (7): e68902. https://doi.org/10.1371/journal.pone.0068902.

Haralambieva, Iana H., Michael J. Gibson, Richard B. Kennedy, Inna G. Ovsyannikova, Nathaniel D. Warner, Diane E. Grill, and Gregory A. Poland. 2017. "Characterization of Rubella-Specific Humoral Immunity Following Two Doses of MMR Vaccine Using Proteome Microarray Technology." *PLoS ONE* 12 (11): e0188149. https://doi.org/10.1371/journal.pone.0188149.

Hasan, Mohammad Rubayet, Mahbuba Rahman, Taushif Khan, Amira Saeed, Sathyavathi Sundararaju, Annaliza Flores, Phillip Hawken, et al. 2021. "Virome-Wide Serological Profiling Reveals Association of Herpesviruses with Obesity." *Scientific Reports* 11 (1): 2562. https://doi.org/10.1038/s41598-021-82213-4.

Heiss, Kirsten, Jasmin Heidepriem, Nico Fischer, Laura K. Weber, Christine Dahlke, Thomas Jaenisch, and Felix F. Loeffler. 2020. "Rapid Response to Pandemic Threats: Immunogenic Epitope Detection of Pandemic Pathogens for Diagnostics and Vaccine Development Using Peptide Microarrays." *Journal of Proteome Research*, September. https://doi.org/10.1021/acs.jproteome.0c00484.

Hou, Xin, Xiaomei Zhang, Xian Wu, Minya Lu, Dan Wang, Meng Xu, Hongye Wang, et al. 2020. "Serum Protein Profiling Reveals a Landscape of Inflammation and Immune Signaling in Early-Stage COVID-19 Infection." *Molecular & Cellular Proteomics* 19 (11): 1749–59. https://doi.org/10.1074/mcp.RP120.002128.

Hu, Shaohui, Zhi Xie, Jiang Qian, Seth Blackshaw, and Heng Zhu. 2011. "Functional Protein Microarray Technology: Functional Protein Microarray Technology." *Wiley Interdisciplinary Reviews: Systems Biology and Medicine* 3 (3): 255–68. https://doi.org/10.1002/wsbm.118.

Jiang, He-wei, Yang Li, Hai-nan Zhang, Wei Wang, Xiao Yang, Huan Qi, Hua Li, Dong Men, Jie Zhou, and Sheng-ce Tao. 2020. "SARS-CoV-2 Proteome Microarray for Global Profiling of COVID-19 Specific IgG and IgM Responses." *Nature Communications* 11 (1): 1–11.

Katz, Chen, Liron Levy-Beladev, Shahar Rotem-Bamberger, Tiago Rito, Stefan G. D. Rüdiger, and Assaf Friedler. 2011. "Studying Protein–Protein Interactions Using Peptide Arrays." *Chemical Society Reviews* 40 (5): 2131–2145. https://doi.org/10.1039/C0CS00029A.

Kalantari-Dehaghi, Mina, Sookhee Chun, Aziz Alami Chentoufi, Jozelyn Pablo, Li Liang, Gargi Dasgupta, Douglas M. Molina, et al. 2012. "Discovery of Potential Diagnostic and Vaccine Antigens in Herpes Simplex Virus 1 and 2 by Proteome-Wide Antibody Profiling." *Journal of Virology* 86 (8): 4328–39. https://doi.org/10.1128/JVI.05194-11.

Li, Siting, Guang Song, Yina Bai, Ning Song, Jiuliang Zhao, Jian Liu, and Chaojun Hu. 2021. "Applications of Protein Microarrays in Biomarker Discovery for Autoimmune Diseases." *Frontiers in Immunology* 12 (May): 645632. https://doi.org/10.3389/fimmu.2021.645632.

Li, Yang, Ming-liang Ma, Qing Lei, Feng Wang, Wei Hong, Dan-yun Lai, Hongyan Hou, et al. 2021. "Linear Epitope Landscape of the SARS-CoV-2 Spike Protein Constructed from 1,051 COVID-19 Patients." *Cell Reports* 34 (13). https://doi.org/10.1016/j.celrep.2021.108915.

Luevano, Martha, Hans-Ulrich Bernard, Hugo A. Barrera-Saldaña, Victor Trevino, Alejandro Garcia-Carranca, Luisa L. Villa, Bradley J. Monk, et al. 2010. "High-Throughput Profiling of the Humoral Immune Responses against Thirteen Human Papillomavirus Types by Proteome Microarrays." *Virology* 405 (1): 31–40. https://doi.org/10.1016/j.virol.2010.05.011.

Ma, MingLiang, DaWei Shi, Yang Li, Wei Hong, DanYun Lai, JunBiao Xue, et al. 2021. "Systematic Profiling of SARS-CoV-2-specific IgG Responses Elicited by an Inactivated Virus Vaccine Identifies Peptides and Proteins for Predicting Vaccination Efficacy." *Cell Discovery* 7 (1): 67. https://doi.org/10.1038/s41421-021-00309-7.

Mishra, Nischay, Xi Huang, Shreyas Joshi, Cheng Guo, James Ng, Riddhi Thakkar, Yongjian Wu, et al. 2021. "Immunoreactive Peptide Maps of SARS-CoV-2." *Communications Biology* 4 (1): 1–7. https://doi.org/10.1038/s42003-021-01743-9.

Mishra, Pushpendra Mani, Vladimir N. Uversky, and Rajanish Giri. 2018. "Molecular Recognition Features in Zika Virus Proteome." *Journal of Molecular Biology* 430 (16). Intrinsically Disordered Proteins: 2372–88. https://doi.org/10.1016/j.jmb.2017.10.018.

Musicò, Angelo, Roberto Frigerio, Alessandro Mussida, Luisa Barzon, Alessandro Sinigaglia, Silvia Riccetti, Federico Gobbi, et al. 2021. "SARS-CoV-2 Epitope Mapping on Microarrays Highlights Strong Immune-Response to N Protein Region." *Vaccines* 9 (1): 35. https://doi.org/10.3390/vaccines9010035.

Natesan, Mohan, and Robert G. Ulrich. 2010. "Protein Microarrays and Biomarkers of Infectious Disease." *International Journal of Molecular Sciences* 11 (12): 5165–83. https://doi.org/10.3390/ijms11125165.

Qi, Huan, Huiqiong Zhou, Daniel Mark Czajkowsky, Shujuan Guo, Yang Li, Nan Wang, Yi Shi, et al. 2017. "Rapid Production of Virus Protein Microarray Using Protein Microarray Fabrication through Gene Synthesis (PAGES)." *Molecular & Cellular Proteomics* 16 (2): 288–99. https://doi.org/10.1074/mcp.M116.064873.

Romanov, Valentin, S. Nikki Davidoff, Adam R. Miles, David W. Grainger, Bruce K. Gale, and Benjamin D. Brooks. 2014. "A Critical Comparison of Protein Microarray Fabrication Technologies." *The Analyst* 139 (6): 1303–26. https://doi.org/10.1039/C3AN01577G.

Sauer, Ursula. 2017. "Analytical Protein Microarrays: Advancements Towards Clinical Applications." *Sensors* 17 (2): 256. https://doi.org/10.3390/s17020256

Schwarz, Tatjana, Kirsten Heiss, Yuvaraj Mahendran, Fiordiligie Casilag, Florian Kurth, Leif E. Sander, Clemens-Martin Wendtner, et al. 2021. "SARS-CoV-2 Proteome-Wide Analysis Revealed Significant Epitope Signatures in COVID-19 Patients." *Frontiers in Immunology* 12. https://www.frontiersin.org/article/10.3389/fimmu.2021.629185.

Shrock, Ellen, Eric Fujimura, Tomasz Kula, Richard T. Timms, I-Hsiu Lee, Yumei Leng, Matthew L. Robinson, Brandon M. Sie, Mamie Z. Li, Yuezhou Chen, Jennifer Logue, Adam Zuiani, Denise McCulloch, Felipe J. N. Lelis, Stephanie Henson, Daniel R. Monaco, Meghan Travers, Shaghayegh Habibi, William A. Clarke, Patrizio Caturegli, Oliver Laeyendecker, Alicja Piechocka-Trocha, Jonathan Z. Li, Ashok Khatri, Helen Y. Chu, MGH COVID-19 Collection & Processing Team16‡, et al. 2020. "Viral Epitope Profiling of COVID-19 Patients Reveals Cross-Reactivity and Correlates of Severity." *Science* 370 (6520): eabd4250. https://doi.org/10.1126/science.abd4250.

Sutandy, F. X. Reymond, Jiang Qian, Chien-Sheng Chen, and Heng Zhu. 2013. "Overview of Protein Microarrays." *Current Protocols in Protein Science* 72 (1). https://doi.org/10.1002/0471140864.ps2701s72.

Uzoma, Ijeoma, and Heng Zhu. 2013. "Interactome Mapping: Using Protein Microarray Technology to Reconstruct Diverse Protein Networks." *Genomics, Proteomics & Bioinformatics* 11 (1): 18–28. https://doi.org/10.1016/j.gpb.2012.12.005.

Wang, Hongye, Xian Wu, Xiaomei Zhang, Xin Hou, Te Liang, Dan Wang, Fei Teng, et al. 2020. "SARS-CoV-2 Proteome Microarray for Mapping COVID-19 Antibody Interactions at Amino Acid Resolution." *ACS Central Science* 6 (12): 2238–49. https://doi.org/10.1021/acscentsci.0c00742.

Zhang, Xiaomei, Mei Zheng, Te Liang, Haijian Zhou, Hongye Wang, et al. 2021. "Inhibitor Screening of Spike Variants Reveals the Heterogeneity of Neutralizing Antibodies to COVID-19 Infection and Vaccination." *medRxiv*. https://doi.org/10.1101/2021.05.15.21257254.

Zhao, Jia, Tasneem H. Patwa, Manoj Pal, Weilian Qiu, and David M. Lubman. 2009. "Analysis of Protein Glycosylation and Phosphorylation Using Liquid Phase Separation, Protein Microarray Technology, and Mass Spectrometry." In *Mass Spectrometry of Proteins and Peptides*, edited by Mary S. Lipton and Ljiljana Paša-Tolic, 492:321–51. Methods In Molecular Biology. Totowa, NJ: Humana Press. https://doi.org/10.1007/978-1-59745-493-3_20.

7

COVID-19 pathogenesis and host immune response

Surbhi Bihani, Shalini Aggarwal, and Arup Acharjee

CONTENTS

List of abbreviations

ACE2 – Angiotensin-converting enzyme 2
ALI – Acute lung injury
APP – Acute-phase protein
ARDS – Acute respiratory distress syndrome
ARI – Acute respiratory infections
BALF – Bronchoalveolar lavage fluid
CNCB-NGDC – National Genomics Data Center-China National Center for Bio-information
CRP – C-reactive protein
DC – Dendritic cells
DDX5 – DEAD-box RNA helicase 5
DMVs – Double-membrane vesicles
ER – Endoplasmic reticulum
IFN – Interferon
IL – Interleukin
IRFs – Interferon regulatory factors
ISG – Interferon-stimulated genes
MAVS – Mitochondrial antiviral signaling protein

DOI: 10.1201/9781003220787-7

MDA-5 – Melanoma differentiation-associated protein 5
MERS-CoV – Middle East respiratory syndrome coronavirus
MHC – Major histocompatibility complex
Mpro – Main protease
mTOR – Mammalian target of rapamycin
MyD88 – Myeloid differentiation primary response 88
NF-κB – Nuclear factor kappa-light-chain-enhancer of activated B cells
NKRF – Nuclear factor-kappa B repressing factor
NLRs – Nucleotide oligomerization domain-like receptors
NOD – Nucleotide oligomerization domain
Nsp – Non-structural proteins
ORF – Open reading frame
PAMP – Pathogen-associated molecular patterns
PBMC – Peripheral blood mononuclear cell
PLpro – Papain-like protease
PPI – Protein–protein interaction
PRR – Pattern recognition receptors
RBD – Receptor-binding domain
RdRp – RNA-dependent RNA polymerase
RIG-I – Retinoic acid-inducible gene I
RLRs – Retinoic acid-inducible gene I-like receptors
ROS – Reactive oxygen species
SAA – Serum amyloid A
SARS-CoV – Severe acute respiratory syndrome coronavirus
sgRNA – Subgenomic RNAs
TCR – T-cell receptor
TLRs – Toll-like receptors
TRIF – TIR-domain-containing adaptor inducing interferon β

7.1 Introduction

Pathogenesis by an infecting organism is a victory of pathogen infectivity on the targeted host. It is often precipitated by the insufficiency of the host immune system to tackle the invading pathogen. Host–pathogen interactions lead to the dysregulation of various host biological pathways, while the host activates the innate immune system that essentially results in inflammation leading to symptoms, progression, and severity of the disease. Pathogen tries to evade the innate immune system and faces the adaptive immune system of the host. Depending on the location of the infection, the host exhibits localized lesions, systemic effects, or a combination of both. Host response to the entry, immune escape, and viral replication leads to different disease severity levels. Hence, to understand pathogenesis, one needs to understand the functions of viral proteins and, simultaneously, the host's response. The host's genetic makeup, predisposed conditions/comorbidities, recovery capacity, age, gender, geographic location, and many other factors decide the host's capability to fight disease. This involves the different modes of SARS-CoV-2 virus entry in the host system, host response against the virus, the role of the viral proteins in pathogenesis, and the omics approaches that have facilitated the better resolution of the host–pathogen interactions and the host response.

7.1.1 General course of RNA virus infection

Acute respiratory infections (ARI) are among the top five causes of mortality worldwide and respiratory virus infections are significant contributors to that (Legand et al. 2013), causing an enormous burden to the healthcare systems and the socioeconomic conditions. The emerging respiratory viruses such as influenza viruses and coronaviruses, among others, present a continual threat to global health and the economy (Tyrrell, Allen, and Carson 2017). To initiate any response against an invading pathogen, the host has to detect the presence of the pathogen. Pathogen-associated molecular patterns (PAMPs) are distinct molecular structures associated with different classes of pathogens that the host recognizes through pattern recognition receptors (PRRs). Upon sensing the PAMPs, PRRs initiate downstream signaling to trigger the production of mediators that aid in alerting the immune system, propagating the immune response, protecting the neighboring cells, and clearing the infection. Three classes of PRRs, viz., Toll-like receptors (TLRs), a retinoic acid-inducible gene I (RIG-I)-like receptors (RLRs), and nucleotide oligomerization domain (NOD)-like receptors (NLRs), are currently known to be involved in sensing of viral components like ssRNA, dsRNA, RNA with 5′-triphosphate ends, and genomic DNA (Takeuchi and Akira 2009). Endoplasmic reticulum (ER) stress induced by virus infection and fusion of viral envelope with cell membrane can also trigger host immune response (Hrincius et al. 2015; Holm et al. 2012).

Recognition of virus-associated PAMPs by PRRs initiates downstream signaling cascades that lead to activation of primarily two classes of transcription factors—nuclear factor kappa-light-chain-enhancer of activated B cells (NF-κB) and interferon regulatory factors (IRFs) (Troy and Bosco 2016). These transcription factors act predominantly to induce the expression of cytokines, particularly chemokines and interferons that mediate the antiviral program. Chemokines orchestrate the recruitment of immune cells at the site of infection, while interferons induce the expression of several interferon-stimulated genes (ISGs) that give rise to a potent antiviral state, aid in viral clearance, and prevent viral spread (Duan and Mukherjee 2016; Honda and Taniguchi 2006). Successive waves of cytokines and chemokines are set in, resulting in coordinated and sequential recruitment of neutrophils, cytotoxic T cells, and natural killer cells at early time points, followed by memory T cells, and finally naive T and B lymphocytes (B et al. 2006). The cytokine expression also results in maturation and trafficking of antigen-presenting dendritic cells (DCs) to draining lymph nodes required to activate adaptive immune response (Kohlmeier and Woodland 2009). This coordinated set of reactions starting from detecting the virus to the expression of chemokines and interferons resulting in induction and recruitment of immune cells is fundamental to tackling respiratory virus infections. The early innate response characterized by the release of cytokines restricts the replication of the virus, induces a potent antiviral state, and activates adaptive immunity. The adaptive immune system takes time to respond because it requires selecting virus-specific cells from a humongous pool of naive T and B lymphocytes followed by their proliferation and differentiation into effector and memory cells. Viruses have evolved several mechanisms to evade the immune system and bring about the associated pathology. It is reasonable to think that any virus-associated disease, such as COVID-19, is the repercussion of the concerned virus trying to trick and evade the immune system.

7.1.2 SARS-CoV-2: Entry and life cycle

SARS-CoV-2 and severe acute respiratory syndrome coronavirus (SARS-CoV) and Middle East respiratory syndrome coronavirus (MERS-CoV) are placed under the *Betacoronavirus* genus of the Coronaviridae family. The members of Coronaviridae are enveloped with positive-sense, single-stranded RNA viruses (Gorbalenya et al. 2020). Deep sequencing methods have identified that the full-length genomic RNA of SARS-CoV-2 is around ~30 kb long and comprises 14 open reading frames (Kumar 2020). The spike protein enables cellular entry of the virus by interacting with the host cell surface receptor, angiotensin-converting enzyme 2 (ACE2) (Wang et al. 2020). After the receptor-binding domain (RBD) of spike protein comes in close contact with host ACE2, it is subjected to proteolytic priming by a transmembrane serine protease—TMPRSS2—of the host, which is to enable efficient viral–host membrane fusion reaction (Hoffmann et al. 2020). Alternatively, receptor-mediated endocytosis of the viral particle followed by proteolytic priming of S protein by endosomal cathepsin L is another way of viral entry into the host cell (Hoffmann et al. 2020). Once the viral genome is released inside the host cell, ORF1a and ORF1ab are translated into polyproteins and are cleaved into nsps by host and viral proteases (nsp3 and nsp5 harbor papain-like and 3C-like protease domains, respectively) (Harrison, Lin, and Wang 2020). Nsp12 possesses the main RNA-dependent RNA polymerase activity and, with other nsps, forms the Replicase to enable replication of genomic and subgenomic RNAs (sgRNA) (Harrison, Lin, and Wang 2020). Structural and accessory proteins enable the assembly and egress of progeny virions and aid in viral replication and pathogenesis (Yoshimoto 2020). The viral entry and life cycle are summarized in Figure 7.1.

7.1.3 Recognition of SARS-CoV-2 infection

The host cells detect the entry of the virus by various sensors called PRRs that recognize virus-specific PAMPs. Viral PAMPs are essentially recognized by PRRs like TLR3, TLR4 TLR7, TLR8, RIG-I, and melanoma differentiation-associated protein 5 (MDA-5, which is a part of the RLR family) (Sallenave and Guillot 2020). TLR4 is a cell surface receptor that is likely to interact with viral glycoproteins like the spike protein (Choudhury and Mukherjee 2020). On the other hand, TLR3, TLR7, and TLR8 are localized in the endosomes and engage with single-stranded and double-stranded RNA after endocytosis of viral particles (Lester and Li 2014). Furthermore, RIG-I and MDA-5 are cytoplasmic helicases that can detect the genomic RNA of the virus (Chen et al. 2020). The

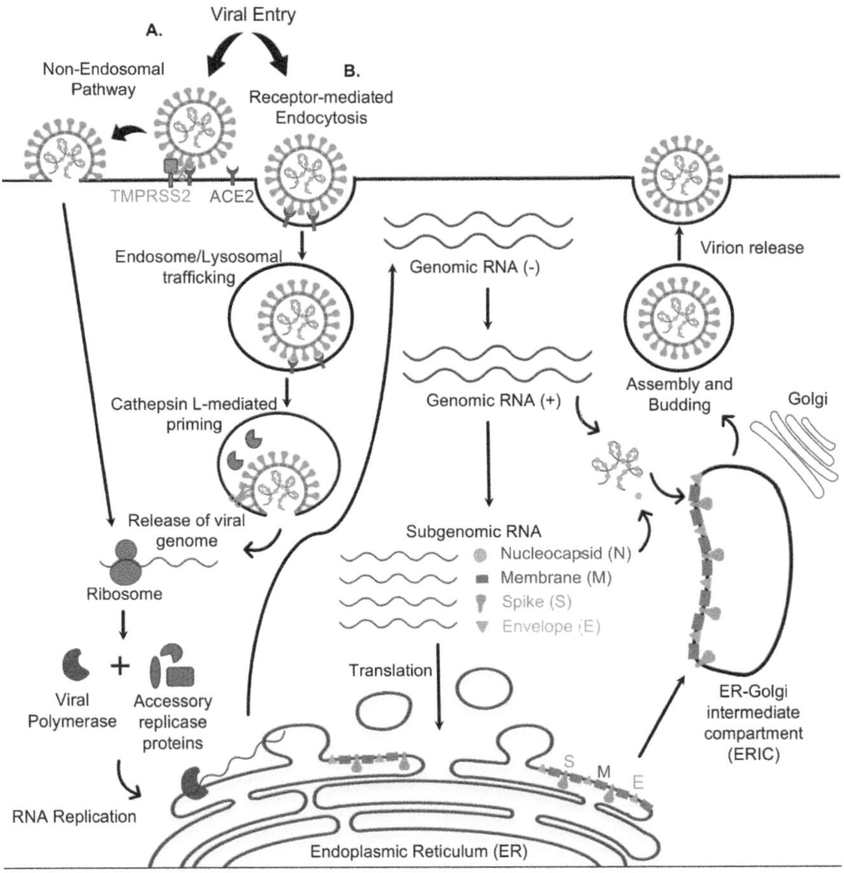

FIGURE 7.1 The course of SARS-CoV2 entry using a different mode of entry: A. TMPRSS2 receptor-mediated non-endosomal pathway; B. ACE-2 receptor-mediated endosomal pathway.

Source: Adapted from Harrison et al. 2020.

engagement of these receptors with the viral products leads to the recruitment of various adaptor proteins, initiating signal transduction pathways that lead to the expression of a variety of immunomodulatory mediators. TLRs signal through myeloid differentiation primary response 88 (MyD88) dependent (TLR4, TLR7, and TLR8) and TIR-domain-containing adaptor inducing interferon β (TRIF or TRICAM1) dependent (TLR3 and TLR4) pathways, while RIG-I and MDA5 activate mitochondrial antiviral signaling protein (MAVS) dependent pathway (Lester and Li 2014; J. Wu et al. 2021). These signaling pathways converge on transcription factors like NF-κB and AP-1, which induce the expression of inflammatory cytokines, and IRFs (like IRF3 and IRF7) that induce interferon response (Park and Iwasaki 2020; Uehata and Takeuchi 2020). The viral sensing mechanism is summarized in Figure 7.2.

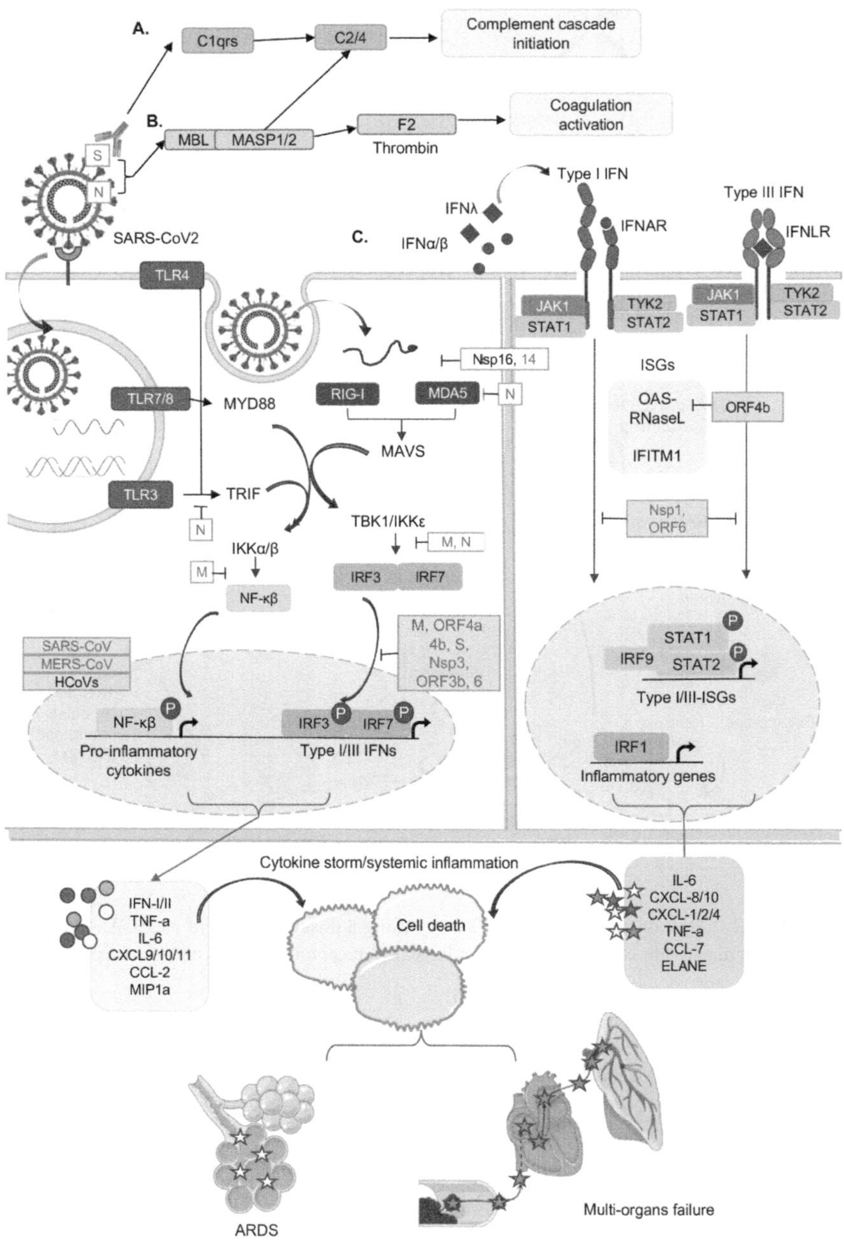

FIGURE 7.2 The activation of the innate response to viral entry leading to A.: activation of complement cascade by classical and lectin pathway, B.: activation of coagulation pathway as a result of thrombin production from lectin pathway, and C.: upregulation of inflammatory biomolecules causing systemic cytokine storm. The systemic cytokine storm results in multiple organ damage along with generating symptoms like respiratory distress due to ARDS.

Source: Adapted from Park and Iwasaki 2020.

7.2 Host response to SARS-CoV-2 infection

The first-line defense of a host against pathogen attack is the innate immune system. Mounting evidence suggests that a disproportionate innate immune response is primarily responsible for the genesis of SARS-CoV-2 infections (Z. Zhou et al. 2020; Blanco-Melo et al. 2020; Hadjadj et al. 2020). A balanced cytokine response accounts for the major first line defense against viruses and is imperative for an effective antiviral response. Heightened release of pro-inflammatory cytokines, also called a cytokine storm, is a common reaction to respiratory viral infections leading to an acute inflammatory state that begins at the infection site and can spread to other parts of the body through systemic circulation. A repercussion of cytokine storm is acute lung injury (ALI) characterized by loss of endothelial and epithelial integrity and excessive transepithelial migration of inflammatory neutrophils (Tisoncik et al. 2012), which is also a principal clinical characteristic of severe COVID-19 patients (L. Li et al. 2020; Ciabattini et al. 2020). Serological profile and bronchoalveolar lavage fluid (BALF) of clinical COVID-19 patients has revealed significantly increased levels of cytokines like IL-6, CCL8, CXCL8 (IL-8), CXCL9, CXCL16, IL-1RA, CCL2, and CXCL2 (Blanco-Melo et al. 2020; Z. Zhou et al. 2020; Liao et al. 2020). These cytokines have varied roles in inflammation and immune reactions that bring about the virus-associated pathophysiology. The cellular landscape of BALF from severe and critical cases of COVID-19 has revealed a higher number of neutrophils and macrophages as compared to moderate cases, and evidence suggests that extensive infiltration of inflammatory immune cells contributes to lung damage and mortality (Barnes et al. 2020; Liao et al. 2020; Merad and Martin 2020). The cytokine storm can also result in multiple organ damage leading to various clinical manifestations and worsening the existing clinical condition in the infected patient. For example, islet cells get damaged due to the cytokine storm, which might accelerate the development of diabetic conditions, and the severe condition is engendered in the case of diabetic patients (Hayden 2020). Hence, comorbidities play a vital role in the conversion of the COVID-19 non-severe cases to severe cases.

Interleukin 6 (IL-6), which has been widely associated with the severity of COVID-19 (J. Zhang et al. 2020), is a pleiotropic cytokine with the ability to induce the production of acute-phase proteins (APP) by liver cells and promote antibody production, T-cell differentiation, and megakaryocyte maturation (Tanaka, Narazaki, and Kishimoto 2014). Furthermore, IL-6 inhibits the production of albumin (responsible for maintaining oncotic pressure of the blood) and induces the production of hepcidin (reduces iron levels in serum by blocking iron transporter—ferroportin—in enterocytes and macrophages) and ferritin (iron-storing blood protein) (Tanaka, Narazaki, and Kishimoto 2014; Edeas, Saleh, and Peyssonnaux 2020). Many of these biological effects of IL-6 corroborate very well with the clinical features of COVID-19. Proteome analysis of sera from severe patients has revealed significant upregulation of APPs like serum amyloid A (SAA), C-reactive protein (CRP), alpha-1-antichymotrypsin, and fibrinogen (D'Alessandro et al. 2020; Shen et al. 2020), suggestive of a hyper-inflammatory state associated with the disease. Coagulation-related abnormalities are a significant risk factor in susceptible patients and can be partly attributed to increased platelet production due to IL-6-induced megakaryocyte maturation (Roncati, Manenti et al. 2020; Biswas and Khan 2020; Al-Samkari et al. 2020; Wool and Miller 2021; Roncati, Ligabue et al. 2020). Markedly decreased serum levels of albumin and iron, also called hypoalbuminemia and hypoferremia, respectively, and high serum levels of hepcidin and ferritin have been associated with disease severity, poor clinical outcomes, and mortality (J. Huang et al. n.d.; Yafei Zhang et al. 2020; Zhao et al. 2020; J. Xia et al. 2019; Nai et al. 2021; Cheng et al. 2020). Excess intracellular iron resulting from high hepcidin levels react with intracellular oxygen, leading to reactive oxygen species (ROS) generation. This may lead to oxidative stress and cellular damage in multiple organs. Interestingly, altered iron homeostasis has been implicated in oxidative cell death called ferroptosis (Ursini and Maiorino 2020), which is crucially involved in neurological disturbances (J.-X. Ren et al. 2020), anosmia (loss of smell) (Dinc et al. 2016), and ageusia (loss of taste) (Osaki et al. 1996), features very commonly manifested in COVID-19.

On the other hand, mounting evidence suggests profoundly poor interferon response in COVID-19 patients. Interferons, classified as type I (IFN α and ß), type II (IFN γ) and type III (IFN λ), are crucial for mounting an effective antiviral response. In addition to inhibiting various viral replication stages, they have several immunomodulatory functions that ultimately act to recruit immune cells and promote adaptive and innate immune responses (Samuel 2001; Biron 1998). While a small subset of interferon-stimulated genes (ISGs) is found to be upregulated in multiple studies, little to no interferon response was observed in sera (Blanco-Melo et al. 2020), peripheral white blood cells (Hadjadj et al. 2020), and BALF (Z. Zhou et al. 2020) of COVID-19 patients. The upregulation of ISGs can be explained by the fact that viral PAMPs can directly induce the expression of some of the ISGs without the involvement of IFNs (Sen 2001). On the contrary, Galani et al. and Broggi et al. observed type I and type III interferons in severe and critical cases (Galani et al. 2021; Broggi et al. 2020), suggesting a delayed response. This might as well be responsible for worsening the situation as delayed production of type I and type III IFNs has been shown to impair lung epithelial barriers in the mice model (Major et al. 2020; Broggi et al. 2020). Thus, the pathogenesis of SARS-CoV-2 infection can be viewed as a vicious cycle of systemic events occurring due to heightened cytokine expression and diminished and/or delayed interferon response.

7.3 COVID-19 and multi-omics

Any biological system is a hugely interconnected and complex system. For decades we have tried to break the understanding of the *in vivo* environment using genomics, transcriptomics, metabolomics, proteomics, and many other branches of science dealing with epigenetics, posttranslational modifications, and many more. However, to understand pathology or disease mechanics better, one needs to integrate these omics studies. The researchers have used various biospecimens and performed multi-omics to integrate the findings facilitating a better understanding of the pathology of SARS-CoV-2 in the human system.

In COVID-19 affected patients, change in the immunological profile includes nutrient drop in blood profile and the rise in inflammatory molecules that diversify with disease progression (Shen et al. 2020). The inflammatory response alteration includes cytokine IL-6 and a rise in the T-cell population resulting in lymphocyte proliferation exhaustion (Zheng et al. 2020). Keratin-19 has been reported to be upregulated in damaged muscle tissues and is suspected to be a potential biomarker for tissue damage (Su et al. 2020). In the patient cohort transforming from moderately severe to severe, it is reported to find further alterations in the immune response–related cellular population such as lymphopenia (X. Cao 2020), robust HLA class II downregulation on monocytes (Wilk et al. 2020), spontaneous rise in inflammatory cytokines levels (Del Valle et al. 2020), altered mammalian target of rapamycin (mTOR) signaling in dendritic cells (Arunachalam et al. 2020), and altered levels of myeloid cells in severe COVID-19 (Schulte-Schrepping et al. 2020; Silvin et al. 2020). Virus infection inside a host requires various biomolecules for its various stages of survival and reproduction, such as invasion, replication, encapsulation, and reinfection of other host cells (Bley, Schöbel, and Herker 2020). This results in alteration of carbon metabolism (D. Wu et al. 2020), nitrogen metabolism (Thomas et al. 2020; D. Wu et al. 2020; Heer et al. 2020), pyrimidine metabolism (D. Wu et al. 2020), lipid metabolism (Shen et al. 2020; J.-W. Song et al. 2020; Maras et al. 2020), and amino acid metabolism (Shen et al. 2020; J.-W. Song et al. 2020; Maras et al. 2020; Thomas et al. 2020) pathways of the host system.

Multi-omics analysis using meta-analysis also revealed the alteration of androgen and estrogen metabolism, amino sugar metabolism, carbon metabolism, glycerophospholipids metabolism, amino acid metabolism, nitrogen metabolism, and vitamin metabolism (Aggarwal et al. 2021). Multi-omics analysis and correlation of infectious diseases such as malaria, tuberculosis, and other deadly infectious diseases enable to understand the commonly altered pathways. The commonly altered pathways observed are RNA splicing, RNA metabolic processes, immune response, and other pathways involved in the host's defense mechanism (Barh et al. 2020). These pathways and target molecules can be beneficial in the repurposing of FDA-approved drugs.

7.4 Possible mechanisms of immune evasion

Gordon et al. sought to characterize the host–virus protein–protein interactions (PPIs) to lay the groundwork for dissecting the exact molecular mechanism of viral pathogenesis (Gordon et al. 2020). The virus undertakes the deception in two crucial acts. Suppression of immune response is act one. The induction of IFN-I is critical for the innate immune response against viral infections (Stetson and Medzhitov 2006). Using biotin-streptavidin affinity purification mass spectrometry (AP-MS), it was found that Orf9b of SARS-CoV-2 localizes on the outer membrane of host cell mitochondrial protein TOM70 to suppress type I interferon (IFN-I) responses (Jiang, Zhang et al. 2020). Further viral proteins ORF6, ORF8, and nucleocapsid also inhibit IFN-I signaling allowing immune evasion (Flower et al. 2021; J.-Y. Li et al. 2020). Moreover, the non-structural protein-1 has been shown to interact with the 40S ribosomal subunit to inhibit the mRNA synthesis of the retinoic acid–inducible gene I-dependent innate immune responses. Therefore, it could entirely block the translation of IFNs and IFN-stimulated genes (ISGs) that usually facilitate viral clearance (Thoms et al. 2020). The upregulation of inflammatory response synergistically complements the inhibition of interferon response.

Act two is the persistent inhibition of the adaptive immune response. It usually takes a couple of weeks to produce antibodies against the virus. PBMC proteome of mild and severe COVID-19 patients revealed a breakdown of T-cell function in severe COVID-19 patients compared to moderately symptomatic cases. T-cell receptor subunits, T-cell surface molecules, MHC class II molecules, T-cell migration stimulators, and TCR signaling kinases are significantly reduced during severe COVID-19. Moreover, B cell antibody secretion is impaired during severe COVID-19, as implied by reduced immunoglobulin subunits expression in these patients. Therefore, more severe COVID-19 patients suffer a functional decline in adaptive immunity, which also corresponded with their more inadequate T-cell and B-cell populations (Liang et al. 2020). This two-act drama of immune response avoidance by the virus allows the virus to spread from one unwary host to another. The different viral proteins and their roles in pathogenesis and the viral life cycle have been summarized in Table 7.1.

TABLE 7.1

SARS-CoV-2 Proteins and Their Role in the Pathogenesis

SARS-CoV-2 Proteins	Role in the Viral Life Cycle and Pathogenesis	Reference
Nsp1	Interacts with 40S ribosomal subunit and inhibits host translation Suppresses interferon signaling and thus plays role in immune evasion Promotes viral gene expression and suppresses host gene expression	Schubert et al. 2020; Lei et al. 2020
Nsp2	Likely to interact with prohibitin 1 and prohibitin 2, host proteins that are involved in several cellular processes such as mitochondrial biogenesis, cell cycle progression, cell differentiation, and apoptosis. This interaction can potentially be responsible for disturbing host intracellular signaling and milieu	Yoshimoto 2020
Nsp3	It is a papain-like protease (PLpro) that cleaves nsp1, nsp2, and nsp3 from the viral polypeptide It plays a role in immune evasion by reversing posttranslational modifications of host proteins (possess deubiquitinating and deISGylating activities). It plays a role in the inhibition of interferon-β response Along with nsp4, it induces membrane rearrangement resulting in the formation of double-membrane vesicles (DMVs). DMVs serve to anchor viral replication and transcription complexes	Freitas et al. 2020; Lei et al. 2020; Hagemeijer et al. 2014
Nsp4	nsp4 interacts with nsp3 and is involved in the formation of virus-induced double-membrane vesicles (DMVs) which are essential for viral replication	Hagemeijer et al. 2014; Sakai et al. 2017

Nsp5	nsp5 is the main protease (Mpro, also known as 3C-like protease) that, after autolytic cleavage from pp1a and pp1ab, cleaves at least 11 sites in the viral polyprotein	Jin et al. 2020
Nsp6	It is a multi-pass membrane protein that works along with nsp3 and nsp4 to induce double-membrane vesicles in infected cells. It is also involved in restricting autophagosome expansion and interferes with the delivery of viral components to lysosomes SARS-CoV-2 nsp6 suppresses IFN-I by binding to TBK1 and blocking STAT1 phosphorylation	Cottam, Whelband, and Wileman 2014; Yoshimoto 2020; H. Xia et al. 2020
Nsp7	Along with nsp8, nsp7 acts as a cofactor for the RNA-dependent RNA polymerase activity of nsp12	Kirchdoerfer and Ward 2019
Nsp8	Along with nsp7, nsp8 acts as a cofactor for the RNA-dependent RNA polymerase activity of nsp12	Kirchdoerfer and Ward 2019
Nsp9	nsp9 is an RNA-binding protein. It is likely to interact with DEAD-box RNA helicase 5 (DDX5) host protein and promote viral replication	Yoshimoto 2020
Nsp10	Stimulates the methyltransferase activities of nsp14 and nsp16 Interacts with host nuclear factor-kappa B-repressing factor (NKRF) to induce the production of IL-8 (chemotactic factor for neutrophils)	Kumar 2020; Yoshimoto 2020; J. Li et al. 2020
Nsp11	Independent function, if any, has not been described yet	Suryawanshi et al. 2021
Nsp12	nsp12 is the RNA-dependent RNA polymerase (RdRp) with the ability to perform both replication and transcription of the viral genome. The nucleoside analog Remdesivir inhibits nsp12	Kumar 2020
Nsp13	It has helicase activity which acts to unwind duplex RNA. The helicase activity is enhanced by nsp12 It is also involved in the capping of the viral mRNA It is likely to interact with host TBK1 and TBKBP1 to inhibit type 1 IFN response	Kumar 2020; Yoshimoto 2020
Nsp14	nsp14 possesses 3′–5′ exoribonuclease (proofreading during RNA replication) and N7-guanine methyltransferase (viral mRNA capping) activities. Interacts with nsp10	Kumar 2020; Yoshimoto 2020
Nsp15	nsp15 is an endoribonuclease that cleaves RNA at the 3′-ends of uridylates. It plays a role in immune evasion as it specifically degrades viral RNA to escape from the host viral sensing mechanism It inhibits type 1 IFN response by binding to host RING finger protein 41 and neuregulin receptor degradation protein 1 (Nrdp1, an E3 ubiquitin-protein ligase). Binding to Nrdp1 supports and aggravates TLR signaling-dependent production of pro-inflammatory cytokines	Kumar 2020
Nsp16	nsp16 forms a complex with nsp10 It is involved in viral mRNA capping. The 2′-*O*-ribose-methyltransferase activity of nsp16 methylates the first transcribed nucleotide—adenine—of the viral RNA	Kumar 2020; Yoshimoto 2020

(Continued)

TABLE 7.1 (Continued)

SARS-CoV-2 Proteins	Role in the Viral Life Cycle and Pathogenesis	Reference
Spike (S) protein	The spike glycoprotein is essential for cellular entry of the virus. It interacts with the host ACE2 receptor	Y. Huang et al. 2020
ORF3a	ORF3a is likely to stimulate the production of pro-inflammatory cytokines via interaction with TRIM59 (E3 ubiquitin ligase) protein. TRIM59 is a negative regulator of NF-κB and IRF signaling It has an apoptosis-inducing property which is relatively lesser than ORF3a of SARS-CoV. This might be responsible for the mild and/or asymptomatic infection during the early stages, allowing SARS-CoV-2 to spread more widely. ORF3a of SARS-CoV is known to induce apoptosis by TRAF3- and ASC-mediated activation of NLRP3 inflammasome	Kumar 2020; Siu et al. 2019; Kondo, Watanabe, and Hatakeyama 2012; Y. Ren et al. 2020; Gordon et al. 2020
Envelope protein	It is an integral structural protein involved in viral assembly, egress, and pathogenesis. It can oligomerize and form viroporin The envelope protein of SARS-CoV-2 shares high sequence similarity with that of SARS-CoV. SARS-CoV E protein interacts with host protein—syntenin—to trigger the overproduction of pro-inflammatory cytokines via the p38 MAPK pathway It has been reported to interact with a host protein associated with tight junctions, PALS1, and can be responsible for disrupting the epithelial barrier in alveoli The ion channel (IC) activity of E protein viroporin activates NLRP3-inflammasome leading to the production of IL-1 β. The IC activity of viroporin also contributes toward pulmonary edema and lung damage	Yoshimoto 2020; Kumar 2020; Chan et al. 2020; Jimenez-Guardeño et al. 2014; Schoeman and Fielding 2020; Nieto-Torres et al. 2014
Membrane protein	It is a major structural protein and is highly immunogenic. It is involved in the formation of capsid via interaction with N protein. It aids in the assembly and budding of virus particles In SARS-CoV, it has been shown to induce apoptosis by disrupting PDK1-PKB/Akt interaction	Mahtarin et al. 2020; Yoshimoto 2020; Kumar 2020
ORF6	Interacts with Nsp8 to promote RNA polymerase activity It antagonizes IFNI response—interacts with an importin to prevent nuclear localization of IRF3 and ISGF3, leading to inhibition of IFN response	H. Xia et al. 2020; Yoshimoto 2020
ORF7a	Type 1 transmembrane protein and an accessory protein. It is likely to interact with host immune cells like T-lymphocytes and macrophages. Antagonizes IFN-I response by inhibiting STAT2 phosphorylation	Nizamudeen et al. 2021; Yoshimoto 2020; Z. Cao et al. 2021
ORF7b	It is a type III integral transmembrane protein that localizes in the Golgi apparatus. It is incorporated in viral particles and may help in viral assembly	Kumar 2020; Yoshimoto 2020; Schaecher, Mackenzie, and Pekosz 2007

ORF8	Shown to interact with and downregulate the expression of class I MHC molecules through autophagy-dependent lysosomal degradation. Thus, it helps the virus to evade immune surveillance by preventing the elimination of virus-infected cells by cytotoxic T cells (CTLs) Known to suppress the innate immune response by acting as an IFN antagonist ORF8 of SARS-CoV-2 has the lowest sequence similarity to that of SARS-CoV and can be responsible for some of the differences in the pathogenesis of the two viruses	Yiwen Zhang et al. 2020; J.-Y. Li et al. 2020; Chan et al. 2020
ORF9b	It has been shown to inhibit type I interferon response by associating with TOM70. TOM70 is a mitochondrial adaptor protein involved in virus-mediated induction of IFN response Interferes with RIG-I-MAVS viral-sensing pathway and subsequent production of IFNs. Targets NEMO (NF-κB Essential Modulator) Involved in host–viral interplay as antibodies against ORF9b have been reported in many COVID-19 patients' sera	Jiang, Zhang et al. 2020; Jiang, Li et al. 2020; Lin, Paz, and Hiscott 2010; J. Wu et al. 2021; Gordon et al. 2020
ORF9c	ORF9c has been reported to upregulate IL-1, IL-6, and p38 MAPK signaling pathways and downregulate interferon, complement system, and antigen presentation-associated pathways	Andres et al. 2020
ORF10	Non-essential	Pancer et al. 2020
Nucleocapsid protein	It is immunogenic and a well-conserved multifunctional protein across the coronavirus family. It is primarily involved in the packaging of the viral genome and viral assembly. It also aids in viral RNA replication It regulates the cell cycle by inhibiting the cyclin-CDK complex and thereby inducing hypo-phosphorylation of retinoblastoma (Rb) protein. This prevents S-phase progression SARS-CoV-2 nucleocapsid protein has been shown to suppress cellular RNAi-mediated antiviral immunity It plays a role in innate immunity suppression by inhibiting type 1 interferon response and NF-κB pathway	J.-Y. Li et al. 2020; Zeng et al. 2020; McBride, van Zyl, and Fielding 2014; Mu et al. 2020; Yoshimoto 2020

7.5 Conclusions and future perspectives

National Genomics Data Center-China National Center for Bio-information (CNCB-NGDC) (CNCB-NGDC Members and Partners 2021; S. Song et al. 2020) is preparing an online platform for COVID-19 research to make the published findings available to the researchers, clinicians, and industries. In addition to understanding the current situation, the pertinent situations to be looked into are the long-term effect of COVID-19 on the recovered population, relapse, new variants, and therapeutic strategy for the long run. For reasons as yet unknown, a subset of the population, the effect of COVID-19 is far from over after the viral clearance (Mahase 2020). These patients continue to suffer the debilitating effects of COVID-19, much after being declared free of the virus itself, thus affecting their typical day-to-day functioning. A study on 201 patients suggests that nearly 70% had perturbations in one or more organs four months after their initial symptoms of SARS-CoV-2 infection (Iacobucci 2020; Carfì et al. 2020). Multi-omic studies on long COVID-19 patients and long-term longitudinal studies on completely recovered versus "long COVID" cohorts are the need of the hour to deepen our knowledge of this phenomenon.

SARS-CoV-2 has a proclivity to mutate. A cursory glance at the Nextstrain portal (Hadfield et al. 2018), which is driven by data obtained from the GISAID repository (Shu and McCauley 2017), will reveal the extraordinary breadth of mutations the virus has acquired in its short lifespan leading to 12 clades. Using genomic data available with GISAID (Shu and McCauley 2017), 42,177 amino acid replacements, 175 coding region insertions, and 1,277 coding region deletions have been detected in the SARS-CoV-2 genome concerning the Wuhan-Hu-1 isolate (NC_045512.2) using CoV-Glue (Singer et al. 2020). Further, the virus's ability to infect other mammals has led to the spillover of the infection to unsuspecting animals, which might reinfect the human host and lead to further mutations (P. Zhou and Shi 2021). All of this is a recipe for yet another wave of this amaranthine pandemic. The most common fallout of recurrent mutation is the rise of neutralizing antibody-escape mutants. Neutralizing antibodies are essential to prevent reinfection as well as viral clearance. However, because SARS-CoV-2 has spread far and wide in the human population with heterogeneous immune response and has been subjected to selection pressures, convalescent plasma therapy might dupe the existing antibodies by the guise of mutant strains. The spike protein, which associates with host ACE2 protein during infection, is the most prone to mutations (Weisblum et al. 2020). *In vitro* experiments have also shown that this is a definite possibility (Baum et al. 2020). There have been reports where reinfections have occurred due to phylogenetically distinct strains (Larson et al. 2020; To et al. 2020; Van Elslande et al. 2020). With vaccination drives ongoing worldwide, the challenge of a neutralizing antibody-escape mutant raising its head is omnipresent, and vigilance is crucial.

REFERENCES

Aggarwal, Shalini, Arup Acharjee, Amrita Mukherjee, Mark S. Baker, and Sanjeeva Srivastava. 2021. "Role of Multiomics Data to Understand Host—Pathogen Interactions in COVID-19 Pathogenesis." *Journal of Proteome Research*, January. https://doi.org/10.1021/acs.jproteome.0c00771.

Al-Samkari, Hanny, Rebecca S. Karp Leaf, Walter H. Dzik, Jonathan C. T. Carlson, Annemarie E. Fogerty, Anem Waheed, Katayoon Goodarzi, et al. 2020. "COVID-19 and Coagulation: Bleeding and Thrombotic Manifestations of SARS-CoV-2 Infection." *Blood* 136 (4): 489–500. https://doi.org/10.1182/blood.2020006520.

Andres, Ana Dominguez, Yongmei Feng, Alexandre Rosa Campos, Jun Yin, Chih-Cheng Yang, Brian James, Rabi Murad, et al. 2020. "SARS-CoV-2 ORF9c Is a Membrane-Associated Protein That Suppresses Antiviral Responses in Cells." *BioRxiv*, August. https://doi.org/10.1101/2020.08.18.256776.

Arunachalam, Prabhu S., Florian Wimmers, Chris Ka Pun Mok, Ranawaka A. P. M. Perera, Madeleine Scott, Thomas Hagan, Natalia Sigal, et al. 2020. "Systems Biological Assessment of Immunity to Mild versus Severe COVID-19 Infection in Humans." *Science (New York, N.Y.)* 369 (6508): 1210–20. https://doi.org/10.1126/science.abc6261.

Barh, Debmalya, Sandeep Tiwari, Marianna E. Weener, Vasco Azevedo, Aristóteles Góes-Neto, M. Michael Gromiha, and Preetam Ghosh. 2020. "Multi-Omics-Based Identification of SARS-CoV-2 Infection Biology and Candidate Drugs against COVID-19." *Computers in Biology and Medicine* 126 (November): 104051. https://doi.org/10.1016/j.compbiomed.2020.104051.

Barnes, Betsy J., Jose M. Adrover, Amelia Baxter-Stoltzfus, Alain Borczuk, Jonathan Cools-Lartigue, James M. Crawford, Juliane Daßler-Plenker, et al. 2020. "Targeting Potential Drivers of COVID-19: Neutrophil Extracellular Traps." *Journal of Experimental Medicine* 217 (e20200652). https://doi.org/10.1084/jem.20200652.

Baum, Alina, Benjamin O. Fulton, Elzbieta Wloga, Richard Copin, Kristen E. Pascal, Vincenzo Russo, Stephanie Giordano, et al. 2020. "Antibody Cocktail to SARS-CoV-2 Spike Protein Prevents Rapid Mutational Escape Seen with Individual Antibodies." *Science* 369 (6506): 1014–18. https://doi.org/10.1126/science.abd0831.

Biron, Christine A. 1998. "Role of Early Cytokines, Including Alpha and Beta Interferons (IFN-A\β), in Innate and Adaptive Immune Responses to Viral Infections." *Seminars in Immunology* 10 (5): 383–90. https://doi.org/10.1006/smim.1998.0138.

Biswas, Indranil, and Gausal A. Khan. 2020. "Coagulation Disorders in COVID-19: Role of Toll-like Receptors." *Journal of Inflammation Research*. Dove Press, October 29, 2020. https://doi.org/10.2147/JIR.S271768.

Blanco-Melo, Daniel, Benjamin E. Nilsson-Payant, Wen-Chun Liu, Skyler Uhl, Daisy Hoagland, Rasmus Møller, Tristan X. Jordan, et al. 2020. "Imbalanced Host Response to SARS-CoV-2 Drives Development of COVID-19." *Cell* 181 (5): 1036–45.e9. https://doi.org/10.1016/j.cell.2020.04.026.

Bley, Hanna, Anja Schöbel, and Eva Herker. 2020. "Whole Lotta Lipids—From HCV RNA Replication to the Mature Viral Particle." *International Journal of Molecular Sciences* 21 (8): 2888. https://doi.org/10.3390/ijms21082888.

Broggi, Achille, Sreya Ghosh, Benedetta Sposito, Roberto Spreafico, Fabio Balzarini, Antonino Lo Cascio, Nicola Clementi, et al. 2020. "Type III Interferons Disrupt the Lung Epithelial Barrier upon Viral Recognition." *Science* 369 (6504): 706–12. https://doi.org/10.1126/science.abc3545.

Cao, Xuetao. 2020. "COVID-19: Immunopathology and Its Implications for Therapy." *Nature Reviews Immunology* 20 (5): 269–70. https://doi.org/10.1038/s41577-020-0308-3.

Cao, Zengguo, Hongjie Xia, Ricardo Rajsbaum, Xianzhu Xia, Hualei Wang, and Pei-Yong Shi. 2021. "Ubiquitination of SARS-CoV-2 ORF7a Promotes Antagonism of Interferon Response." *Cellular & Molecular Immunology* 18 (3): 746–48. https://doi.org/10.1038/s41423-020-00603-6.

Carfì, Angelo, Roberto Bernabei, Francesco Landi, and for the Gemelli Against COVID-19 Post-Acute Care Study Group. 2020. "Persistent Symptoms in Patients After Acute COVID-19." *JAMA* 324 (6): 603. https://doi.org/10.1001/jama.2020.12603.

Chan, Jasper Fuk-Woo, Kin-Hang Kok, Zheng Zhu, Hin Chu, Kelvin Kai-Wang To, Shuofeng Yuan, and Kwok-Yung Yuen. 2020. "Genomic Characterization of the 2019 Novel Human-Pathogenic Coronavirus Isolated from a Patient with Atypical Pneumonia after Visiting Wuhan." *Emerging Microbes & Infections* 9 (1): 221–36. https://doi.org/10.1080/22221751.2020.1719902.

Chen, Yan-Mei, Yuanting Zheng, Ying Yu, Yunzhi Wang, Qingxia Huang, Feng Qian, Lei Sun, et al. 2020. "Blood Molecular Markers Associated with COVID-19 Immunopathology and Multi-Organ Damage." *The EMBO Journal* 39 (24): e105896. https://doi.org/10.15252/embj.2020105896.

Cheng, Linlin, Haolong Li, Liubing Li, Chenxi Liu, Songxin Yan, Haizhen Chen, and Yongzhe Li. 2020. "Ferritin in the Coronavirus Disease 2019 (COVID-19): A Systematic Review and Meta-Analysis." *Journal of Clinical Laboratory Analysis* 34 (10): e23618. https://doi.org/10.1002/jcla.23618.

Choudhury, Abhigyan, and Suprabhat Mukherjee. 2020. "In Silico Studies on the Comparative Characterization of the Interactions of SARS-CoV-2 Spike Glycoprotein with ACE-2 Receptor Homologs and Human TLRs." *Journal of Medical Virology* 92 (10): 2105–13. https://doi.org/10.1002/jmv.25987.

Ciabattini, Annalisa, Paolo Garagnani, Francesco Santoro, Rino Rappuoli, Claudio Franceschi, and Donata Medaglini. 2020. "Shelter from the Cytokine Storm: Pitfalls and Prospects in the Development of SARS-CoV-2 Vaccines for an Elderly Population." *Seminars in Immunopathology*, November: 1–16. https://doi.org/10.1007/s00281-020-00821-0.

CNCB-NGDC Members and Partners. 2021. "Database Resources of the National Genomics Data Center, China National Center for Bioinformation in 2021." *Nucleic Acids Research* 49 (D1): D18–28. https://doi.org/10.1093/nar/gkaa1022.

Cottam, Eleanor M., Matthew C. Whelband, and Thomas Wileman. 2014. "Coronavirus NSP6 Restricts Autophagosome Expansion." *Autophagy* 10 (8): 1426–41. https://doi.org/10.4161/auto.29309.

D'Alessandro, Angelo, Tiffany Thomas, Monika Dzieciatkowska, Ryan C. Hill, Richard O. Francis, Krystalyn E. Hudson, James C. Zimring, Eldad A. Hod, Steven L. Spitalnik, and Kirk C. Hansen. 2020. "Serum Proteomics in COVID-19 Patients: Altered Coagulation and Complement Status as a Function of IL-6 Level." *Journal of Proteome Research* 19 (11): 4417–27. https://doi.org/10.1021/acs.jproteome.0c00365.

Del Valle, Diane Marie, Seunghee Kim-Schulze, Hsin-Hui Huang, Noam D. Beckmann, Sharon Nirenberg, Bo Wang, Yonit Lavin, et al. 2020. "An Inflammatory Cytokine Signature Predicts COVID-19 Severity and Survival." *Nature Medicine* 26 (10): 1636–43. https://doi.org/10.1038/s41591-020-1051-9.

Dinc, Mehmet Emre, Abdullah Dalgic, Seckin Ulusoy, Denizhan Dizdar, Omer Develioglu, and Murat Topak. 2016. "Does Iron Deficiency Anemia Affect Olfactory Function?" *Acta Oto-Laryngologica* 136 (7): 754–57. https://doi.org/10.3109/00016489.2016.1146410.

Duan, Linna, and Eric Mukherjee. 2016. "Janeway's Immunobiology, Ninth Edition." *The Yale Journal of Biology and Medicine* 89 (3): 424–25.

Edeas, Marvin, Jumana Saleh, and Carole Peyssonnaux. 2020. "Iron: Innocent Bystander or Vicious Culprit in COVID-19 Pathogenesis?" *International Journal of Infectious Diseases* 97 (August): 303–05. https://doi.org/10.1016/j.ijid.2020.05.110.

Flower, Thomas G., Cosmo Z. Buffalo, Richard M. Hooy, Marc Allaire, Xuefeng Ren, and James H. Hurley. 2021. "Structure of SARS-CoV-2 ORF8, a Rapidly Evolving Immune Evasion Protein." *Proceedings of the National Academy of Sciences* 118 (2). https://doi.org/10.1073/pnas.2021785118.

Freitas, Brendan T., Ian A. Durie, Jackelyn Murray, Jaron E. Longo, Holden C. Miller, David Crich, Robert Jeff Hogan, Ralph A. Tripp, and Scott D. Pegan. 2020. "Characterization and Noncovalent Inhibition of the Deubiquitinase and DeISGylase Activity of SARS-CoV-2 Papain-Like Protease." *ACS Infectious Diseases* 6 (8): 2099–109. https://doi.org/10.1021/acsinfecdis.0c00168.

Galani, Ioanna-Evdokia, Nikoletta Rovina, Vicky Lampropoulou, Vasiliki Triantafyllia, Maria Manioudaki, Eleftherios Pavlos, Evangelia Koukaki, et al. 2021. "Untuned Antiviral Immunity in COVID-19 Revealed by Temporal Type I/III Interferon Patterns and Flu Comparison." *Nature Immunology* 22 (1): 32–40. https://doi.org/10.1038/s41590-020-00840-x.

Gorbalenya, Alexander E., Susan C. Baker, Ralph S. Baric, Raoul J. de Groot, Christian Drosten, Anastasia A. Gulyaeva, Bart L. Haagmans, et al. 2020. "The Species Severe Acute Respiratory Syndrome-Related Coronavirus : Classifying 2019-NCoV and Naming It SARS-CoV-2." *Nature Microbiology* 5 (4): 536–44. https://doi.org/10.1038/s41564-020-0695-z.

Gordon, David E., Gwendolyn M. Jang, Mehdi Bouhaddou, Jiewei Xu, Kirsten Obernier, Kris M. White, Matthew J. O'Meara, et al. 2020. "A SARS-CoV-2 Protein Interaction Map Reveals Targets for Drug Repurposing." *Nature* 583 (7816): 459–68. https://doi.org/10.1038/s41586-020-2286-9.

Hadfield, James, Colin Megill, Sidney M. Bell, John Huddleston, Barney Potter, Charlton Callender, Pavel Sagulenko, Trevor Bedford, and Richard A. Neher. 2018. "Nextstrain: Real-Time Tracking of Pathogen Evolution." *Bioinformatics* 34 (23): 4121–23. https://doi.org/10.1093/bioinformatics/bty407.

Hadjadj, Jérôme, Nader Yatim, Laura Barnabei, Aurélien Corneau, Jeremy Boussier, Nikaïa Smith, Hélène Péré, et al. 2020. "Impaired Type I Interferon Activity and Inflammatory Responses in Severe COVID-19 Patients." *Science* 369 (6504): 718–24. https://doi.org/10.1126/science.abc6027.

Hagemeijer, Marne C., Iryna Monastyrska, Janice Griffith, Peter van der Sluijs, Jarno Voortman, Paul M. van Bergen en Henegouwen, Annelotte M. Vonk, Peter J. M. Rottier, Fulvio Reggiori, and Cornelis A. M. de Haan. 2014. "Membrane Rearrangements Mediated by Coronavirus Nonstructural Proteins 3 and 4." *Virology* 458–459 (June): 125–35. https://doi.org/10.1016/j.virol.2014.04.027.

Harrison, Andrew G., Tao Lin, and Penghua Wang. 2020. "Mechanisms of SARS-CoV-2 Transmission and Pathogenesis." *Trends in Immunology* 41 (12): 1100–15. https://doi.org/10.1016/j.it.2020.10.004.

Hayden, Melvin R. 2020. "An Immediate and Long-Term Complication of COVID-19 May Be Type 2 Diabetes Mellitus: The Central Role of β-Cell Dysfunction, Apoptosis and Exploration of Possible Mechanisms." *Cells* 9 (11). https://doi.org/10.3390/cells9112475.

Heer, Collin D., Daniel J. Sanderson, Lynden S. Voth, Yousef M. O. Alhammad, Mark S. Schmidt, Samuel A. J. Trammell, Stanley Perlman, Michael S. Cohen, Anthony R. Fehr, and Charles Brenner. 2020. "Coronavirus Infection and PARP Expression Dysregulate the NAD Metabolome: An Actionable Component of Innate Immunity." *Journal of Biological Chemistry*, October. https://doi.org/10.1074/jbc.RA120.015138.

Hoffmann, Markus, Hannah Kleine-Weber, Simon Schroeder, Nadine Krüger, Tanja Herrler, Sandra Erichsen, Tobias S. Schiergens, et al. 2020. "SARS-CoV-2 Cell Entry Depends on ACE2 and TMPRSS2 and Is Blocked by a Clinically Proven Protease Inhibitor." *Cell* 181 (2): 271–80.e8. https://doi.org/10.1016/j.cell.2020.02.052.

Holm, Christian K., Søren B. Jensen, Martin R. Jakobsen, Natalia Cheshenko, Kristy A. Horan, Hanne B. Moeller, Regina Gonzalez-Dosal, et al. 2012. "Virus-Cell Fusion as a Trigger of Innate Immunity Dependent on the Adaptor STING." *Nature Immunology* 13 (8): 737–43. https://doi.org/10.1038/ni.2350.

Honda, Kenya, and Tadatsugu Taniguchi. 2006. "IRFs: Master Regulators of Signalling by Toll-like Receptors and Cytosolic Pattern-Recognition Receptors." *Nature Reviews. Immunology* 6 (9): 644–58. https://doi.org/10.1038/nri1900.

Hrincius, Eike R., Swantje Liedmann, David Finkelstein, Peter Vogel, Shane Gansebom, Amali E. Samarasinghe, Dahui You, Stephania A. Cormier, and Jonathan A. McCullers. 2015. "Acute Lung Injury Results from Innate Sensing of Viruses by an ER Stress Pathway." *Cell Reports* 11 (10): 1591–603. https://doi.org/10.1016/j.celrep.2015.05.012.

Huang, Jiaofeng, Aiguo Cheng, Rahul Kumar, Yingying Fang, Gongping Chen, Yueyong Zhu, and Su Lin. n.d. "Hypoalbuminemia Predicts the Outcome of COVID-19 Independent of Age and Co-morbidity." *Journal of Medical Virology*. Accessed January 18, 2021. https://doi.org/10.1002/jmv.26003.

Huang, Yuan, Chan Yang, Xin-feng Xu, Wei Xu, and Shu-wen Liu. 2020. "Structural and Functional Properties of SARS-CoV-2 Spike Protein: Potential Antivirus Drug Development for COVID-19." *Acta Pharmacologica Sinica* 41 (9): 1141–49. https://doi.org/10.1038/s41401-020-0485-4.

Iacobucci, Gareth. 2020. "Long Covid: Damage to Multiple Organs Presents in Young, Low Risk Patients." *BMJ* 371 (November). https://doi.org/10.1136/bmj.m4470.

Jiang, He-wei, Yang Li, Hai-nan Zhang, Wei Wang, Xiao Yang, Huan Qi, Hua Li, Dong Men, Jie Zhou, and Sheng-ce Tao. 2020. "SARS-CoV-2 Proteome Microarray for Global Profiling of COVID-19 Specific IgG and IgM Responses." *Nature Communications* 11 (1): 3581. https://doi.org/10.1038/s41467-020-17488-8.

Jiang, He-wei, Hai-nan Zhang, Qing-feng Meng, Jia Xie, Yang Li, Hong Chen, Yun-xiao Zheng, et al. 2020. "SARS-CoV-2 Orf9b Suppresses Type I Interferon Responses by Targeting TOM70." *Cellular & Molecular Immunology* 17 (9): 998–1000. https://doi.org/10.1038/s41423-020-0514-8.

Jimenez-Guardeño, Jose M., Jose L. Nieto-Torres, Marta L. DeDiego, Jose A. Regla-Nava, Raul Fernandez-Delgado, Carlos Castaño-Rodriguez, and Luis Enjuanes. 2014. "The PDZ-Binding Motif of Severe Acute Respiratory Syndrome Coronavirus Envelope Protein Is a Determinant of Viral Pathogenesis." *PLOS Pathogens* 10 (8): e1004320. https://doi.org/10.1371/journal.ppat.1004320.

Jin, Zhenming, Xiaoyu Du, Yechun Xu, Yongqiang Deng, Meiqin Liu, Yao Zhao, Bing Zhang, et al. 2020. "Structure of M Pro from SARS-CoV-2 and Discovery of Its Inhibitors." *Nature* 582 (7811): 289–93. https://doi.org/10.1038/s41586-020-2223-y.

Kirchdoerfer, Robert N., and Andrew B. Ward. 2019. "Structure of the SARS-CoV Nsp12 Polymerase Bound to Nsp7 and Nsp8 Co-Factors." *Nature Communications* 10 (1): 2342. https://doi.org/10.1038/s41467-019-10280-3.

Kohlmeier, Jacob E., and David L. Woodland. 2009. "Immunity to Respiratory Viruses." *Annual Review of Immunology* 27 (1): 61–82. https://doi.org/10.1146/annurev.immunol.021908.132625.

Kondo, T., M. Watanabe, and S. Hatakeyama. 2012. "TRIM59 Interacts with ECSIT and Negatively Regulates NF-KB and IRF-3/7-Mediated Signal Pathways." *Biochemical and Biophysical Research Communications* 422 (3): 501–07. https://doi.org/10.1016/j.bbrc.2012.05.028.

Kumar, V. 2020. "Understanding the Complexities of SARS-CoV2 Infection and Its Immunology: A Road to Immune-Based Therapeutics." *International Immunopharmacology* 88 (November): 106980. https://doi.org/10.1016/j.intimp.2020.106980.

Larson, Derek, Sterling L. Brodniak, Logan J. Voegtly, Regina Z. Cer, Lindsay A. Glang, Francisco J. Malagon, Kyle A. Long, et al. 2020. "A Case of Early Reinfection With Severe Acute Respiratory Syndrome Coronavirus 2 (SARS-CoV-2)." *Clinical Infectious Diseases*, no. ciaa1436 (September). https://doi.org/10.1093/cid/ciaa1436.

Legand, Anaïs, Sylvie Briand, Nikki Shindo, W. Abdullah Brooks, Menno D. de Jong, Jeremy Farrar, Ximena Aguilera, and Frederick G. Hayden. 2013. "Addressing the Public Health Burden of Respiratory Viruses: The Battle against Respiratory Viruses (BRaVe) Initiative." *Future Virology* 8 (10): 953–68. https://doi.org/10.2217/fvl.13.85.

Lei, Xiaobo, Xiaojing Dong, Ruiyi Ma, Wenjing Wang, Xia Xiao, Zhongqin Tian, Conghui Wang, et al. 2020. "Activation and Evasion of Type I Interferon Responses by SARS-CoV-2." *Nature Communications* 11 (1): 3810. https://doi.org/10.1038/s41467-020-17665-9.

Lester, Sandra N., and Kui Li. 2014. "Toll-Like Receptors in Antiviral Innate Immunity." *Journal of Molecular Biology*, Antiviral Innate Immunity (Part II), 426 (6): 1246–64. https://doi.org/10.1016/j.jmb.2013.11.024.

Li, Jingjiao, Mingquan Guo, Xiaoxu Tian, Xin Wang, Xing Yang, Ping Wu, Chengrong Liu, et al. 2020. "Virus-Host Interactome and Proteomic Survey Reveal Potential Virulence Factors Influencing SARS-CoV-2 Pathogenesis." *Med*, July. https://doi.org/10.1016/j.medj.2020.07.002.

Li, Jin-Yan, Ce-Heng Liao, Qiong Wang, Yong-Jun Tan, Rui Luo, Ye Qiu, and Xing-Yi Ge. 2020. "The ORF6, ORF8 and Nucleocapsid Proteins of SARS-CoV-2 Inhibit Type I Interferon Signaling Pathway." *Virus Research* 286 (September): 198074. https://doi.org/10.1016/j.virusres.2020.198074.

Li, Liyang, Qihong Huang, Diane C. Wang, David H. Ingbar, and Xiangdong Wang. 2020. "Acute Lung Injury in Patients with COVID-19 Infection." *Clinical and Translational Medicine* 10 (1): 20–27. https://doi.org/10.1002/ctm2.16.

Liang, Qiming, Jingjiao Li, Mingquan Guo, Xiaoxu Tian, Chengrong Liu, Xin Wang, Xing Yang, Ping Wu, Zixuan Xiao, and Yafei Qu. 2020. "Virus-Host Interactome and Proteomic Survey of PMBCs from COVID-19 Patients Reveal Potential Virulence Factors Influencing SARS-CoV-2 Pathogenesis." *BioRxiv*.

Liao, Mingfeng, Yang Liu, Jing Yuan, Yanling Wen, Gang Xu, Juanjuan Zhao, Lin Cheng, et al. 2020. "Single-Cell Landscape of Bronchoalveolar Immune Cells in Patients with COVID-19." *Nature Medicine* 26 (6): 842–44. https://doi.org/10.1038/s41591-020-0901-9.

Lin, Rongtuan, Suzanne Paz, and John Hiscott. 2010. "Tom70 Imports Antiviral Immunity to the Mitochondria." *Cell Research* 20 (9): 971–73. https://doi.org/10.1038/cr.2010.113.

Mahase, Elisabeth. 2020. "Covid-19: What Do We Know about 'Long Covid'?" *BMJ* 370 (July). https://doi.org/10.1136/bmj.m2815.

Mahtarin, Rumana, Shafiqul Islam, Md Jahirul Islam, M. Obayed Ullah, Md Ackas Ali, and Mohammad A. Halim. 2020. "Structure and Dynamics of Membrane Protein in SARS-CoV-2." *Journal of Biomolecular Structure and Dynamics*: 1–14. https://doi.org/10.1080/07391102.2020.1861983.

Major, Jack, Stefania Crotta, Miriam Llorian, Teresa M. McCabe, Hans Henrik Gad, Simon L. Priestnall, Rune Hartmann, and Andreas Wack. 2020. "Type I and III Interferons Disrupt Lung Epithelial Repair during Recovery from Viral Infection." *Science* 369 (6504): 712–17. https://doi.org/10.1126/science.abc2061.

Maras, Jaswinder Singh, Shvetank Sharma, Adil Bhat, Reshu aggrawal, Ekta Gupta, and Shiv K Sarin. 2020. "Multi-Omics Integration Analysis of Respiratory Specimen Characterizes Baseline Molecular Determinants Associated with COVID-19 Diagnosis." *Preprint*. Infectious Diseases (except HIV/AIDS). https://doi.org/10.1101/2020.07.06.20147082.

McBride, Ruth, Marjorie van Zyl, and Burtram C. Fielding. 2014. "The Coronavirus Nucleocapsid Is a Multifunctional Protein." *Viruses* 6 (8): 2991–3018. https://doi.org/10.3390/v6082991.

Merad, Miriam, and Jerome C. Martin. 2020. "Pathological Inflammation in Patients with COVID-19: A Key Role for Monocytes and Macrophages." *Nature Reviews Immunology* 20 (6): 355–62. https://doi.org/10.1038/s41577-020-0331-4.

Mu, Jingfang, Jiuyue Xu, Leike Zhang, Ting Shu, Di Wu, Muhan Huang, Yujie Ren, et al. 2020. "SARS-CoV-2-Encoded Nucleocapsid Protein Acts as a Viral Suppressor of RNA Interference in Cells." *Science China Life Sciences* 63 (9): 1413–16. https://doi.org/10.1007/s11427-020-1692-1.

Nai, Antonella, Nicola Ivan Lorè, Alessia Pagani, Rebecca De Lorenzo, Simona Di Modica, Fabio Saliu, Daniela Maria Cirillo, Patrizia Rovere-Querini, Angelo A. Manfredi, and Laura Silvestri. 2021. "Hepcidin Levels Predict Covid-19 Severity and Mortality in a Cohort of Hospitalized Italian Patients." *American Journal of Hematology* 96 (1): E32–35. https://doi.org/10.1002/ajh.26027.

"Nextstrain/Ncov/Global." n.d. Accessed January 18, 2021. https://nextstrain.org/ncov/global.

Nieto-Torres, Jose L., Marta L. DeDiego, Carmina Verdiá-Báguena, Jose M. Jimenez-Guardeño, Jose A. Regla-Nava, Raul Fernandez-Delgado, Carlos Castaño-Rodriguez, et al. 2014. "Severe Acute Respiratory Syndrome Coronavirus Envelope Protein Ion Channel Activity Promotes Virus Fitness and Pathogenesis." *PLoS Pathogens* 10 (5): e1004077. https://doi.org/10.1371/journal.ppat.1004077.

Nizamudeen, Zubair Ahmed, Emma-Ruoqi Xu, Vivin Karthik, Mohamed Halawa, Kenton P. Arkill, Andrew M. Jackson, David O. Bates, and Jonas Emsley. 2021. "Structural Assessment of SARS-CoV2 Accessory Protein ORF7a Predicts LFA-1 and Mac-1 Binding Potential." *Bioscience Reports* 41 (BSR20203837). https://doi.org/10.1042/BSR20203837.

Osaki, Tokio, Masashi Ohshima, Yukinobu Tomita, Noriko Matsugi, and Yuka Nomura. 1996. "Clinical and Physiological Investigations in Patients with Taste Abnormality." *Journal of Oral Pathology & Medicine* 25 (1): 38–43. https://doi.org/10.1111/j.1600-0714.1996.tb01221.x.

Pancer, Katarzyna, Aleksandra Milewska, Katarzyna Owczarek, Agnieszka Dabrowska, Michał Kowalski, Paweł P. Łabaj, Wojciech Branicki, Marek Sanak, and Krzysztof Pyrc. 2020. "The SARS-CoV-2 ORF10 Is Not Essential in Vitro or in Vivo in Humans." *PLoS Pathogens* 16 (12): e1008959. https://doi.org/10.1371/journal.ppat.1008959.

Park, Annsea, and Akiko Iwasaki. 2020. "Type I and Type III Interferons—Induction, Signaling, Evasion, and Application to Combat COVID-19." *Cell Host & Microbe* 27 (6): 870–78. https://doi.org/10.1016/j.chom.2020.05.008.

Piqueras, B., J. Connolly, H. Freitas, A. K. Palucka, and J. Banchereau. 2006. "Upon Viral Exposure, Myeloid and Plasmacytoid Dendritic Cells Produce 3 Waves of Distinct Chemokines to Recruit Immune Effectors." *Blood*, April 1, 2006. https://doi.org/10.1182/blood-2005-07-2965.

Ren, Jia-Xin, Xin Sun, Xiu-Li Yan, Zhen-Ni Guo, and Yi Yang. 2020. "Ferroptosis in Neurological Diseases." *Frontiers in Cellular Neuroscience* 14. https://doi.org/10.3389/fncel.2020.00218.

Ren, Yujie, Ting Shu, Di Wu, Jingfang Mu, Chong Wang, Muhan Huang, Yang Han, et al. 2020. "The ORF3a Protein of SARS-CoV-2 Induces Apoptosis in Cells." *Cellular & Molecular Immunology* 17 (8): 881–83. https://doi.org/10.1038/s41423-020-0485-9.

Roncati, Luca, Giulia Ligabue, Vincenzo Nasillo, Beatrice Lusenti, William Gennari, Luca Fabbiani, Claudia Malagoli, et al. 2020. "A Proof of Evidence Supporting Abnormal Immunothrombosis in Severe COVID-19: Naked Megakaryocyte Nuclei Increase in the Bone Marrow and Lungs of Critically Ill Patients." *Platelets* 31 (8): 1085–89. https://doi.org/10.1080/09537104.2020.1810224.

Roncati, Luca, Antonio Manenti, Gianrocco Manco, and Alberto Farinetti. 2020. "Abdominal Aortic Thrombosis Complicating COVID-19 Pneumonia." *Annals of Vascular Surgery* 67 (August): 8–9. https://doi.org/10.1016/j.avsg.2020.05.033.

Sakai, Yusuke, Kengo Kawachi, Yutaka Terada, Hiroko Omori, Yoshiharu Matsuura, and Wataru Kamitani. 2017. "Two-Amino Acids Change in the Nsp4 of SARS Coronavirus Abolishes Viral Replication." *Virology* 510 (October): 165–74. https://doi.org/10.1016/j.virol.2017.07.019.

Sallenave, Jean-Michel, and Loïc Guillot. 2020. "Innate Immune Signaling and Proteolytic Pathways in the Resolution or Exacerbation of SARS-CoV-2 in Covid-19: Key Therapeutic Targets?" *Frontiers in Immunology* 11. https://doi.org/10.3389/fimmu.2020.01229.

Samuel, Charles E. 2001. "Antiviral Actions of Interferons." *Clinical Microbiology Reviews* 14 (4): 778–809. https://doi.org/10.1128/CMR.14.4.778-809.2001.

Schaecher, Scott R., Jason M. Mackenzie, and Andrew Pekosz. 2007. "The ORF7b Protein of Severe Acute Respiratory Syndrome Coronavirus (SARS-CoV) Is Expressed in Virus-Infected Cells and Incorporated into SARS-CoV Particles." *Journal of Virology* 81 (2): 718–31. https://doi.org/10.1128/JVI.01691-06.

Schoeman, Dewald, and Burtram C. Fielding. 2020. "Is There a Link Between the Pathogenic Human Coronavirus Envelope Protein and Immunopathology? A Review of the Literature." *Frontiers in Microbiology* 11. https://doi.org/10.3389/fmicb.2020.02086.

Schubert, Katharina, Evangelos D. Karousis, Ahmad Jomaa, Alain Scaiola, Blanca Echeverria, Lukas-Adrian Gurzeler, Marc Leibundgut, Volker Thiel, Oliver Mühlemann, and Nenad Ban. 2020. "SARS-CoV-2 Nsp1 Binds the Ribosomal MRNA Channel to Inhibit Translation." *Nature Structural & Molecular Biology* 27 (10): 959–66. https://doi.org/10.1038/s41594-020-0511-8.

Schulte-Schrepping, Jonas, Nico Reusch, Daniela Paclik, Kevin Baßler, Stephan Schlickeiser, Bowen Zhang, Benjamin Krämer, et al. 2020. "Severe COVID-19 Is Marked by a Dysregulated Myeloid Cell Compartment." *Cell* 182 (6): 1419–40.e23. https://doi.org/10.1016/j.cell.2020.08.001.

Sen, Ganes C. 2001. "Viruses and Interferons." *Annual Review of Microbiology* 55 (1): 255–81. https://doi.org/10.1146/annurev.micro.55.1.255.

Shen, Bo, Xiao Yi, Yaoting Sun, Xiaojie Bi, Juping Du, Chao Zhang, Sheng Quan, et al. 2020. "Proteomic and Metabolomic Characterization of COVID-19 Patient Sera." *Cell* 182 (1): 59–72.e15. https://doi.org/10.1016/j.cell.2020.05.032.

Shu, Yuelong, and John McCauley. 2017. "GISAID: Global Initiative on Sharing All Influenza Data—from Vision to Reality." *Eurosurveillance* 22 (13). https://doi.org/10.2807/1560-7917.ES.2017.22.13.30494.

Silvin, Aymeric, Nicolas Chapuis, Garett Dunsmore, Anne-Gaëlle Goubet, Agathe Dubuisson, Lisa Derosa, Carole Almire, et al. 2020. "Elevated Calprotectin and Abnormal Myeloid Cell Subsets Discriminate Severe from Mild COVID-19." *Cell* 182 (6): 1401–18.e18. https://doi.org/10.1016/j.cell.2020.08.002.

Singer, J., R. Gifford, M. Cotten, and D. Robertson. 2020. "CoV-GLUE: A Web Application for Tracking SARS-CoV-2 Genomic Variation." *Preprints*, June. https://doi.org/10.20944/preprints202006.0225.v1.

Siu, Kam-Leung, Kit-San Yuen, Carlos Castaño-Rodriguez, Zi-Wei Ye, Man-Lung Yeung, Sin-Yee Fung, Shuofeng Yuan, et al. 2019. "Severe Acute Respiratory Syndrome Coronavirus ORF3a Protein Activates the NLRP3 Inflammasome by Promoting TRAF3-Dependent Ubiquitination of ASC." *FASEB Journal: Official Publication of the Federation of American Societies for Experimental Biology* 33 (8): 8865–77. https://doi.org/10.1096/fj.201802418R.

Song, Jin-Wen, Sin Man Lam, Xing Fan, Wen-Jing Cao, Si-Yu Wang, He Tian, Gek Huey Chua, et al. 2020. "Omics-Driven Systems Interrogation of Metabolic Dysregulation in COVID-19 Pathogenesis." *Cell Metabolism* 32 (2): 188–202.e5. https://doi.org/10.1016/j.cmet.2020.06.016.

Song, Shuhui, Lina Ma, Dong Zou, Dongmei Tian, Cuiping Li, Junwei Zhu, Meili Chen, et al. 2020. "The Global Landscape of SARS-CoV-2 Genomes, Variants, and Haplotypes in 2019nCoVR." *Genomics, Proteomics & Bioinformatics*, December. https://doi.org/10.1016/j.gpb.2020.09.001.

Stetson, Daniel B., and Ruslan Medzhitov. 2006. "Type I Interferons in Host Defense." *Immunity* 25 (3): 373–81. https://doi.org/10.1016/j.immuni.2006.08.007.

Su, Yapeng, Daniel Chen, Dan Yuan, Christopher Lausted, Jongchan Choi, Chengzhen L. Dai, Valentin Voillet, et al. 2020. "Multi-Omics Resolves a Sharp Disease-State Shift between Mild and Moderate COVID-19." *Cell* 183 (6): 1479–95.e20. https://doi.org/10.1016/j.cell.2020.10.037.

Suryawanshi, Rahul K., Raghuram Koganti, Alex Agelidis, Chandrashekhar D. Patil, and Deepak Shukla. 2021. "Dysregulation of Cell Signaling by SARS-CoV-2." *Trends in Microbiology* 29 (3): 224–37. https://doi.org/10.1016/j.tim.2020.12.007.

Takeuchi, Osamu, and Shizuo Akira. 2009. "Innate Immunity to Virus Infection." *Immunological Reviews* 227 (1): 75–86. https://doi.org/10.1111/j.1600-065X.2008.00737.x

Tanaka, Toshio, Masashi Narazaki, and Tadamitsu Kishimoto. 2014. "IL-6 in Inflammation, Immunity, and Disease." *Cold Spring Harbor Perspectives in Biology* 6 (10). https://doi.org/10.1101/cshperspect.a016295.

Thomas, Tiffany, Davide Stefanoni, Julie A. Reisz, Travis Nemkov, Lorenzo Bertolone, Richard O. Francis, Krystalyn E. Hudson, et al. 2020. "COVID-19 Infection Alters Kynurenine and Fatty Acid Metabolism, Correlating with IL-6 Levels and Renal Status." *American Society for Clinical Investigation*, July 23, 2020. https://doi.org/10.1172/jci.insight.140327.

Thoms, Matthias, Robert Buschauer, Michael Ameismeier, Lennart Koepke, Timo Denk, Maximilian Hirschenberger, Hanna Kratzat, et al. 2020. "Structural Basis for Translational Shutdown and Immune Evasion by the Nsp1 Protein of SARS-CoV-2." *Science* 369 (6508): 1249–55. https://doi.org/10.1126/science.abc8665.

Tisoncik, Jennifer R., Marcus J. Korth, Cameron P. Simmons, Jeremy Farrar, Thomas R. Martin, and Michael G. Katze. 2012. "Into the Eye of the Cytokine Storm." *Microbiology and Molecular Biology Reviews : MMBR* 76 (1): 16–32. https://doi.org/10.1128/MMBR.05015-11.

To, Kelvin Kai-Wang, Ivan Fan-Ngai Hung, Jonathan Daniel Ip, Allen Wing-Ho Chu, Wan-Mui Chan, Anthony Raymond Tam, Carol Ho-Yan Fong, et al. 2020. "Coronavirus Disease 2019 (COVID-19) Re-Infection by a Phylogenetically Distinct Severe Acute Respiratory Syndrome Coronavirus 2 Strain Confirmed by Whole Genome Sequencing." *Clinical Infectious Diseases*, no. ciaa1275 (August). https://doi.org/10.1093/cid/ciaa1275.

Troy, Niamh M., and Anthony Bosco. 2016. "Respiratory Viral Infections and Host Responses; Insights from Genomics." *Respiratory Research* 17 (1): 1–12. https://doi.org/10.1186/s12931-016-0474-9.

Tyrrell, Carina S. B., John Lee Y. Allen, and Gail Carson. 2017. "Influenza and Other Emerging Respiratory Viruses." *Medicine* 45 (12): 781–87. https://doi.org/10.1016/j.mpmed.2017.09.003.

Uehata, Takuya, and Osamu Takeuchi. 2020. "RNA Recognition and Immunity-Innate Immune Sensing and Its Posttranscriptional Regulation Mechanisms." *Cells* 9 (7). https://doi.org/10.3390/cells9071701.

Ursini, Fulvio, and Matilde Maiorino. 2020. "Lipid Peroxidation and Ferroptosis: The Role of GSH and GPx4." *Free Radical Biology and Medicine* 152 (May): 175–85. https://doi.org/10.1016/j.freeradbiomed.2020.02.027.

Van Elslande, Jan, Pieter Vermeersch, Kris Vandervoort, Tony Wawina-Bokalanga, Bert Vanmechelen, Elke Wollants, Lies Laenen, et al. 2020. "Symptomatic SARS-CoV-2 Reinfection by a Phylogenetically Distinct Strain." *Clinical Infectious Diseases*, no. ciaa1330 (September). https://doi.org/10.1093/cid/ciaa1330.

Wang, Qihui, Yanfang Zhang, Lili Wu, Sheng Niu, Chunli Song, Zengyuan Zhang, Guangwen Lu, et al. 2020. "Structural and Functional Basis of SARS-CoV-2 Entry by Using Human ACE2." *Cell* 181 (4): 894–904.e9. https://doi.org/10.1016/j.cell.2020.03.045.

Weisblum, Yiska, Fabian Schmidt, Fengwen Zhang, Justin DaSilva, Daniel Poston, Julio C. C. Lorenzi, Frauke Muecksch, et al. 2020. "Escape from Neutralizing Antibodies by SARS-CoV-2 Spike Protein Variants." Edited by Mark Marsh, Jos WM van der Meer, and David Montefiore. *ELife* 9 (October): e61312. https://doi.org/10.7554/eLife.61312.

"WHO | Battle against Respiratory Viruses (BRaVe) Initiative." n.d. WHO. World Health Organization. Accessed January 11, 2021. www.who.int/influenza/patient_care/clinical/brave/en/.

Wilk, Aaron J., Arjun Rustagi, Nancy Q. Zhao, Jonasel Roque, Giovanny J. Martínez-Colón, Julia L. McKechnie, Geoffrey T. Ivison, et al. 2020. "A Single-Cell Atlas of the Peripheral Immune Response in Patients with Severe COVID-19." *Nature Medicine* 26 (7): 1070–76. https://doi.org/10.1038/s41591-020-0944-y.

Wool, Geoffrey D., and Jonathan L. Miller. 2021. "The Impact of COVID-19 Disease on Platelets and Coagulation." *Pathobiology* 88 (1): 15–27. https://doi.org/10.1159/000512007.

Wu, Di, Ting Shu, Xiaobo Yang, Jian-Xin Song, Mingliang Zhang, Chengye Yao, Wen Liu, et al. 2020. "Plasma Metabolomic and Lipidomic Alterations Associated with COVID-19." *National Science Review* 7 (7): 1157–68. https://doi.org/10.1093/nsr/nwaa086.

Wu, Jing, Yuheng Shi, Xiaoyan Pan, Shuang Wu, Ruixia Hou, Yong Zhang, Tiansheng Zhong, et al. 2021. "SARS-CoV-2 ORF9b Inhibits RIG-I-MAVS Antiviral Signaling by Interrupting K63-Linked Ubiquitination of NEMO." *Cell Reports* 34 (7): 108761. https://doi.org/10.1016/j.celrep.2021.108761.

Xia, Hongjie, Zengguo Cao, Xuping Xie, Xianwen Zhang, John Yun-Chung Chen, Hualei Wang, Vineet D. Menachery, Ricardo Rajsbaum, and Pei-Yong Shi. 2020. "Evasion of Type I Interferon by SARS-CoV-2." *Cell Reports* 33 (1): 108234. https://doi.org/10.1016/j.celrep.2020.108234.

Xia, Jian-jun, Fei Wang, Xiao-nan Jiang, Ting-ting Jiang, Li-juan Shen, Yue Liu, Da-li You, et al. 2019. "Serum Iron Levels Are an Independent Predictor of In-Hospital Mortality of Critically Ill Patients: A Retrospective, Single-Institution Study." *Journal of International Medical Research* 47 (1): 66–75. https://doi.org/10.1177/0300060518795528.

Yoshimoto, Francis K. 2020. "The Proteins of Severe Acute Respiratory Syndrome Coronavirus-2 (SARS CoV-2 or n-COV19), the Cause of COVID-19." *The Protein Journal* 39 (3): 198–216. https://doi.org/10.1007/s10930-020-09901-4.

Zeng, Weihong, Guangfeng Liu, Huan Ma, Dan Zhao, Yunru Yang, Muziying Liu, Ahmed Mohammed, et al. 2020. "Biochemical Characterization of SARS-CoV-2 Nucleocapsid Protein." *Biochemical and Biophysical Research Communications* 527 (3): 618–23. https://doi.org/10.1016/j.bbrc.2020.04.136.

Zhang, Jing, Yiqun Hao, Wuling Ou, Fei Ming, Gai Liang, Yu Qian, Qian Cai, et al. 2020. "Serum Interleukin-6 Is an Indicator for Severity in 901 Patients with SARS-CoV-2 Infection: A Cohort Study." *Journal of Translational Medicine* 18 (1): 406. https://doi.org/10.1186/s12967-020-02571-x.

Zhang, Yafei, Liang Zheng, Lan Liu, Mengya Zhao, Jun Xiao, and Qiu Zhao. 2020. "Liver Impairment in COVID-19 Patients: A Retrospective Analysis of 115 Cases from a Single Centre in Wuhan City, China." *Liver International* 40 (9): 2095–103. https://doi.org/10.1111/liv.14455.

Zhang, Yiwen, Junsong Zhang, Yingshi Chen, Baohong Luo, Yaochang Yuan, Feng Huang, Tao Yang, et al. 2020. "The ORF8 Protein of SARS-CoV-2 Mediates Immune Evasion through Potently Downregulating MHC-I." *BioRxiv*, May. https://doi.org/10.1101/2020.05.24.111823.

Zhao, Kang, Jucun Huang, Dan Dai, Yuwei Feng, Liming Liu, and Shuke Nie. 2020. "Serum Iron Level as a Potential Predictor of Coronavirus Disease 2019 Severity and Mortality: A Retrospective Study." *Open Forum Infectious Diseases* 7 (ofaa250). https://doi.org/10.1093/ofid/ofaa250.

Zheng, Meijuan, Yong Gao, Gang Wang, Guobin Song, Siyu Liu, Dandan Sun, Yuanhong Xu, and Zhigang Tian. 2020. "Functional Exhaustion of Antiviral Lymphocytes in COVID-19 Patients." *Cellular & Molecular Immunology* 17 (5): 533–35. https://doi.org/10.1038/s41423-020-0402-2.

Zhou, Peng, and Zheng-Li Shi. 2021. "SARS-CoV-2 Spillover Events." *Science* 371 (6525): 120–22. https://doi.org/10.1126/science.abf6097.

Zhou, Zhuo, Lili Ren, Li Zhang, Jiaxin Zhong, Yan Xiao, Zhilong Jia, Li Guo, et al. 2020. "Heightened Innate Immune Responses in the Respiratory Tract of COVID-19 Patients." *Cell Host & Microbe* 27 (6): 883–90.e2. https://doi.org/10.1016/j.chom.2020.04.017.

8

Putative role of multi-omics technologies in the investigation of persistent effects of COVID-19 on vital human organs

Susmita Ghosh, Akanksha Salkar, and Firuza Parikh

CONTENTS

List of abbreviations

AST – Aspartate aminotransferase
CRP – C-reactive protein
CSF – Cerebrospinal fluid
MRI – Magnetic resonance imaging
PBMC – Peripheral blood mononuclear cell
RNA – Ribonucleic acid

DOI: 10.1201/9781003220787-8

8.1 Introduction

SARS-CoV-2, a single-stranded RNA virus belonging to the coronaviridae family, has caused the worldwide pandemic. The average incubation period is five days, and 97.5% of the infected individuals will present symptoms within 11.5 days of infection. It has been reported that approximately 17–35% of the hospitalized patients are treated in ICU. Hypoxemic respiratory failure is the most common cause for admission into ICU (Wiersinga et al. 2020). Viral entry into the host cells is dependent on angiotensin-converting enzyme 2 (ACE2) and transmembrane protease serine type 2 (TMPRSS2) (Hoffmann et al. 2020). The wide distribution of ACE2 and TMPRSS2 receptors in other vital organs such as the heart, liver, and testis make those more susceptible to viral infection (Beyerstedt et al. 2021). In addition to the common clinical manifestations of the disease, complications such as acute kidney injury, gastrointestinal bleeding, arrhythmias, acute liver injury, and cardiomyopathy have been observed (Figure 8.1). Extensive studies have reported acute hepatic failure, acute kidney injury, and myocardial infarction as consequences of the disease severity (Table 8.1). These findings indicate that understanding the extrapulmonary involvement of SARS-CoV-2 is necessary.

Currently, reverse transcription-polymerase chain reaction (RT-PCR) is the most commonly used method for diagnosing COVID-19 infection. Although many rapid antigen tests are available commercially, they are auxiliary to the RT-PCR test. Imaging modalities such as chest radiography, pulmonary US, chest CT, and angiography are also used to diagnose COVID-19 supplementary to the RT-PCR test. Moreover, chest CT has been used to predict the prognosis of COVID-19 (Feng et al. 2020). Furthermore, a significant number of recovered patients may still have some impairment. There is enough evidence regarding the multi-organ involvement of COVID-19. However, the long-term prevalence of these complications is not well studied. With the current number of people who have been affected by COVID-19, the sudden surge in post-COVID-19 illnesses in recovered people may overwhelm the medical system. Therefore, monitoring these patients becomes necessary to prevent such diseases or to facilitate their prognosis (Balachandar et al. 2020). Thus, in this chapter, we review various complications of COVID-19 in different vital organs, probable long-term effects, and the role of omics technology.

8.2 Effects of COVID-19 on vital organs

Recent reports suggest that there is extrapulmonary involvement in patients infected with SARS-CoV-2. This extrapulmonary involvement can be attributed to the localization of ACE2—a cellular receptor for SARS-CoV-2—on different human cells. The transcriptomics-based studies reporting ACE2 and TMPRSS2 expression in the brain, liver, spleen, kidney, bone marrow, thymus, nasopharynx, oral and nasal mucosa, lung, small intestine, stomach, colon, lymph nodes, skin, arteries, and veins indicate the multi systemic nature of SARS-CoV-2 (Figure 8.2). In this section, we summarize the organ-specific involvement of SARS-CoV-2 (Zhang et al. 2020).

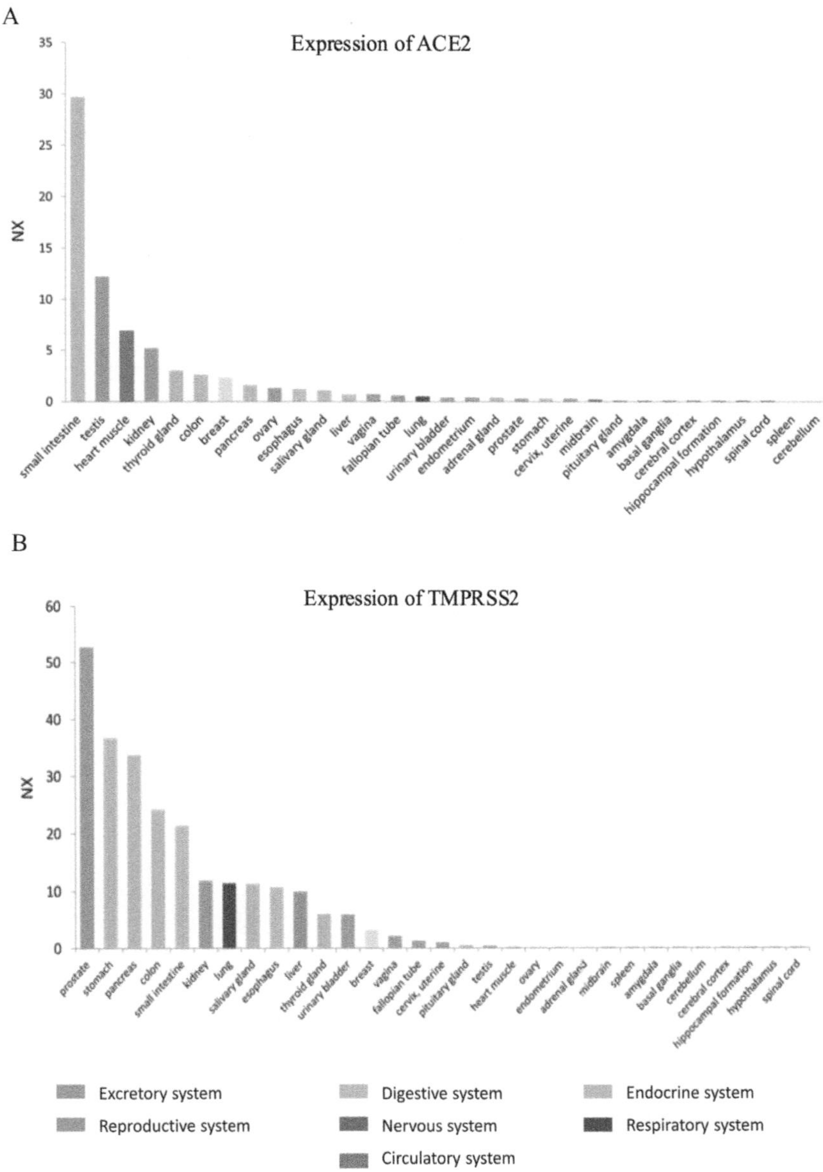

FIGURE 8.1 The gene expression level of two crucial viral receptors (a) ACE2 and (b) TMPRSS2 in human. The y-axis indicates the log10 value of transcripts per million in corresponding organs in x-axis. The expression profile was obtained from GTEX database. https://gtexportal.org/home/

TABLE 8.1

Comorbidities and Clinical Complications of COVID-19

S. No.	Number of Patients (N)	Groups Compared	Median Age	Study Type	Country/ Area/ Region	Existing Comorbidities on Admission	Complications Due to COVID-19	Conclusion
1 Chen et al. 2020	274	Dead versus recovered	62	Retrospective observational study	Shanghai, China	HTN = 93(54+39), DM = 47(24+23) CVD = 23 (16+4) Chronic lung failure = 18 (11+7) Cerebrovascular = 4 (4+0) Chronic kidney disease = 4 (3+1) GI disease = 3(1+2)	ARDS = 196 (113+83) Type I respiratory failure = 18 (18+0) ACI = 89 (72+18), HF = 44 (41+3) Hypoxic encephalopathy = 24 (23+1) AKI = 29(28+1) Disseminated intravascular coagulation = 21 (19+2) ALI = 13 (10+3), 1(1+0)	Preexisting comorbidities to some extent do affect the prognosis of COVID-19. However, regardless of their history, the prevalence of cardiovascular disease, acute cardiac injury, and heart failure was more in deceased patients than in recovered patients
2 Arentz et al. 2020	21	ICU patients	70	Case series	Washington, USA	COPD = 7, congestive heart failure = 9 DM = 7 CKD = 10 ESKD = 2 Obstructive sleep apnea = 6	Coinfection with bacteria =1, virus = 3 Acute kidney failure = 4, cardiomyopathy = 7 Acute hepatic injury = 3 Seizure = 1 Death = 11 Still in hospital = 8 Recovered = 2	High prevalence of cardiomyopathy suggests direct involvement of heart in COVID-19. However, the sample size was too small to conclude anything

3	Richardson et al. 2020	5,700	Discharged, dead, and still in hospital	63	Case series	New York, USA	Cancer = 320, HTN = 3,026, CAD = 595 CHF = 371, Asthma = 479, COPD = 287 Obstructive sleep apnea = 154 Kidney disease: chronic = 268, end-stage = 186, liver disease = 19	AKI = 2,351 (1,869+481) Acute hepatic injury = 56 (3+53)	The number of deaths due to acute kidney injury was more in patients with diabetes than in patients without diabetes
4	Argenziano et al. 2020	1,000	Level of care	63	Case series	New York, USA	HTN = 601, DM = 372, CAD = 131, congestive heart failure = 102, pulmonary disease = 223, asthma = 113, COPD = 66, obstructive sleep apnea = 24, renal disease = 137, history of stroke = 79, HIV = 21, viral hepatitis = 19, cirrhosis = 15	ARDS = 299, AKI = 288, new onset arrhythmia = 79, new onset of heart failure = 24, myocardial infarction = 8* data for 850 patients	Higher rate of incidence of AKI than previous reports. The morbidity and mortality rates were higher. The study provides anticipatory guidance during the ongoing pandemic. Argenziano

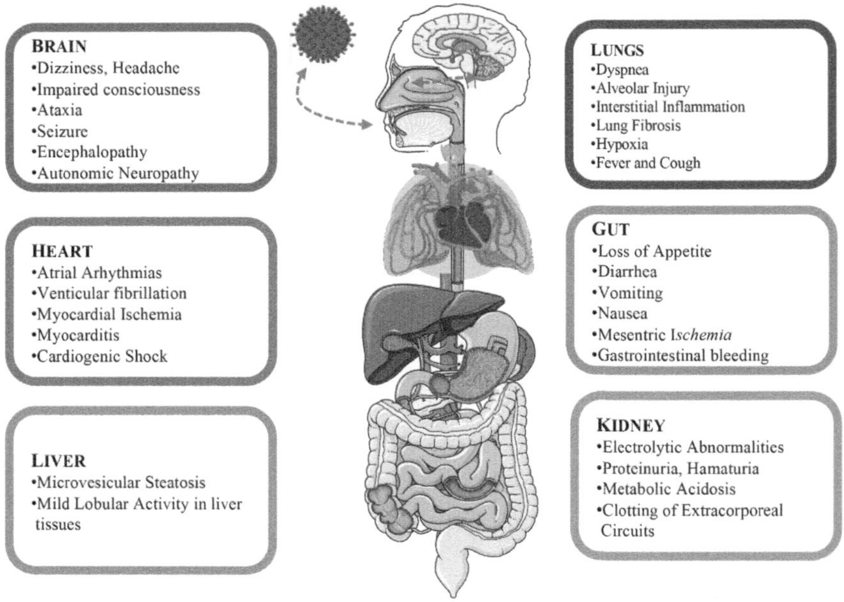

FIGURE 8.2 Overview of extra pulmonary effects of SARS-CoV-2 and the list of organ-specific manifestations.

Source: This figure was created using Servier Medical Art templates, which are licensed under a Creative Commons Attribution 3.0 Unported License; https://smart.servier.com.

8.2.1 Respiratory system

High ACE2 receptor expression in the lungs indicates that the lungs act as a potential proliferation site for the virus. The surface epithelial cells of alveoli, also called pneumocytes, are rich in ACE2, thus allowing the virus to enter, leading to lung function impairment. The lung biopsy report of a patient who died of COVID-19 showed the shedding of pneumocytes, pulmonary edema, and hyaline membrane formation, indicating acute respiratory distress syndrome (Xu et al. 2020). Most characteristic feature of patients' lungs is injury to alveolar epithelial cells with fibroproliferation, capillary damage, bleeding, which leads to pulmonary hypertension or lung fibrosis (Mo et al. 2020). While the virus proliferates and affects the normal physiological function of the lungs, the host immune response to the virus worsens the physiology of the latter. This statement can be explained by the cytokine storm that gag the alveoli and the airway and, thus hinder the gaseous exchange (Prasad and Prasad 2020). Extensive studies on the lung function of SARS-CoV-2 survivors revealed insufficient diffusion capacity, pulmonary dysfunction, and restrictive effects, which directly correlated with disease severity (Torres-Castro et al. 2020). To evaluate the respiratory dysfunction, observations are made based on different lung function tests like pulmonary function test (spirometry), diffusion capacity, lung volume, and chest computed tomography (CT). Mo et al. showed the alteration of the respiratory evaluations in COVID-19 patients at the time of discharge. As per their report, a significant decline of diffusion capacity of carbon monoxide in COVID-19 patients suggests that diffusion membrane is one of the critical responsible factors for lung impairment (Mo et al. 2020). These findings engender a concern about the assessment of the long-term impact of SARS-CoV-2 infection on the lung.

The previous study of SARS survivors had shown various degrees of anomalies in lung function, such as pulmonary function test, diffusion capacity, and impairment of exercise at different time points after disease onset. This study reported a decline in forced expiratory flow, persistent abnormality in diffusion capacity (DLCO) at 24 months post-illness, suggesting the intra-alveolar diffusion pathway's impairment. Besides, the CT scan images strengthened the evidence of functional impairment in the lungs. Opacity and air-trapping in the lungs were two prominent findings in SARS patients' CT images (Ngai et al. 2010). This study gave us an insight into the persistent SARS symptoms in the lungs. An extensive meta-analysis was conducted combining various cohort studies to understand the lung abnormalities in postinfection COVID-19 patients. The study confirmed the prevalence of altered diffusion capacity in a higher percentage of COVID-19 patients after one month of their recovery (Frija-Masson et al. 2020). Intriguingly, one-fourth of the patient cohort in a study conducted by Zhao et al. (2020) were found with gave us an insight into the abnormal pulmonary diffusion in one-fourth of the patient cohort three months after discharge. The elevation of D-dimer, an important prognostic marker for impaired diffusion capacity, draws attention to the proper care of the patients having no severe respiratory symptoms. More recently, Huang et al. (2021) showed the persistence of different pulmonary abnormalities in patients even after six months of disease onset. A considerable proportion of patients were identified with lung diffusion impairment, most of them had disease severity during their active phase of infection. The chest CT report of those patients confirmed the consistency of ground-glass opacity with irregular lines due to this disease. Accounting for all studies reported to date, it is clear that persistance of disease consequences is more in severe patients than non-severe individuals.

8.2.2 Cardiovascular system

The heart is the second organ targeted by the SARS-CoV-2 virus due to the higher expression of ACE2 in the heart muscle. According to scRNA seq data, more than 7.5% of myocardial cells expressed ACE2 (Zou et al. 2020). This renders the cardiomyocytes highly susceptible to attack by SARS-CoV-2. ACE2 is a protector of the vascular system. Intuitively, ACE2 impairments lead to severe cardiac dysfunction, increased atherosclerosis, and endothelial damage. The incidence rate of myocardial injury due to the cardiovascular complication of COVID-19 is 8–12%. Notably, a higher prevalence of cardiovascular diseases and myocardial injury in patients admitted to ICU has been reported (Bansal 2020). Shaha and his colleagues (2020) summarized the putative mechanism of acute cardiac injury in COVID-19 patients. They summarized thatsystemic inflammatory response syndrome (SIRS)/cytokine storm inciting dysregulated immune response, oxygen supply–demand mismatch due to hypoxia, microvascular injury as a result of microvascular thrombi formation, direct cardiotropic myocardial injury, and SARS-CoV-2-induced ACE-mediated damage are potential causes of acute cardiac injury (ACI).

Previously, the SARS-CoV RNA has been detected in 7 out of 20 autopsy heart samples from COVID-19 patients. Moreover, subsequent staining also had demonstrated macrophage infiltration with associated myocardial damage (Oudit et al. 2009). This indicates that SARS-CoV can infect myocardial cells; however, whether these findings can be associated with SARS-CoV-2 remains undetermined as no study till now has reported the presence of SARS-CoV-2 RNA in myocardial cells. Nonetheless, the occurrence of severe myocarditis has been reported in several cases (Agricola et al. 2020). Bradley et al. (2020) have reported that in autopsy samples, the SARS-CoV-2 RNA was detected in myocardiocytes. However, immunohistochemical and electron microscopic findings did not show the presence of RNA. This can be due to contamination by the circulating virus as the person presented with viremia. In addition, immunohistochemistry of all 14 autopsy samples showed interstitial fibrosis followed by myocyte hypertrophy, replacement fibrosis, and myocardial amyloid. Moreover, they observed that cardiac findings were non-specific and associated with preexisting comorbidities (Bradley 2020). In addition, Puntmann et al. (2020) reported abnormal cardiac MR findings in 78 patients who have recently recovered from COVID-19. They reported that these patients presented at least one of the following on CMR: myocardial native T1 ($n=73$), raised myocardial native T2 ($n=60$), myocardial late gadolinium enhancement ($n=32$), or pericardial enhancement ($n=22$).

Troponin is a protein present in the heart muscle that regulates normal heart function. In COVID-19 patients, the troponin levels are often found to be elevated (Wang, Hu et al. 2020). In their study evaluating CMR findings in patients who have recently recovered from COVID-19, Puntmann et al. (2020) have reported that high-sensitivity troponin T (hsTnT) significantly elevated in 5 patients and detectable in 71 patients.

8.2.3 Nervous system

The central and peripheral nervous systems are the other hotspots of ACE2 receptors, making them more susceptible to SARS-CoV-2. Thus, neurological abnormalities are common manifestations observed in 45% of severe patients and 85% of patients having ARDS. The neurological manifestations range from non-specific symptoms like headache, dizziness, and smell loss seen in patients with mild COVID-19 to critical conditions like encephalopathy, ischemic stroke, and brain damage in severe patients.

There is a huge rift in the direct infection of the virus in the brain among researchers; two shreds of evidence confirmed the presence of SARS-CoV-2 in cerebrospinal fluid and temporal lobe. To assess the direct pathogenicity of the virus, Ladecola and his team proposed possible mechanisms of viral entry into the brain deriving from brain invasion strategies of other coronaviruses. The blood–brain barrier is the common route of SARS-CoV-2 for brain invasion, where the olfactory lobe also plays an essential role for the same (Iadecola et al. 2020). Erickson et al. summarized that the inflammatory cytokines secreted in response to viral infection are critical for breaching the blood–brain barrier, often accompanied by comorbidities like preexisting cardiovascular disease or neurological disease (Erickson and Banks 2018). The presence of viral nucleocapsid protein in CD68+ cells and viral RNA in macrophages in the bronchoalveolar lavage samples indicates that infected immune cells are carriers for viral entry into the brain (Bost et al. 2020).

Although the prolonged consequences of COVID-19 on the brain are still not fully understood, a few studies reported acute disseminated encephalomyelitis, acute hemorrhagic encephalopathy, and Guillain–Barre syndrome as postinfectious neurological manifestations in COVID-19 survivors. These complications are characterized as immune-mediated manifestations that include both central and peripheral nervous systems. Therefore, these findings generate a concern to assess the possibilities of various neuropathies and neurodegenerative diseases in COVID-19 recovered patients.

8.2.4 Excretory system

The kidney plays a vital role in filtering out various toxins, waste products, and extra water from our body. Initial reports on the burden of acute kidney injury (AKI) during SARS-CoV-2 infection indicated a moderately low (about 0.5%) burden. However, recent studies reported an incidence going up to 56.9% and is associated with mortality in COVID-19 patients (Cheng et al. 2020). Nadim et al. (2020) reported that AKI affects >50% of ICU patients and >20% of hospitalized patients. Liu et al. (2021) summarized that the AKI in patients with COVID-19 can be caused by viral load–induced cytotoxicity of renal resident cells, kidney hyperfusion due to symptoms like fever, vomit, diarrhea, and shock, the nephrotoxicity of some drugs, and organ crosstalk such as cardiorenal syndrome, hypoxia, and rhabdomyolysis. Moreover, mechanical ventilation and comorbidities such as hypertension, diabetes, chronic cardiology disease, or chronic liver disease can also lead to AKI in patients with COVID-19.

As per Dong et al. (2020), the ACE2 expression in normal kidneys is enriched in proximal tubule cells and it is not expressed in immune cells and glomerular parietal epithelial cells. Suryawanshi et al. (2020) analyzed 14 distinct tissue scRNA-seq datasets and determined that in the kidney, proximal tubular cells and tubular progenitor cells co-expressed ACE2 and TMPRSS2. High ACE2 and TMPRSS2 expression levels in nephron epithelial cells, endothelial cells, and mesangial cells of the kidney were observed from the scRNA-seq data from the GEO dataset (GSE134355) (Figure 8.2). These data showed that both ACE2 and TMPRSS2 were highly expressed in tissues and cells of the kidney. Many studies have presented evidence for prominent acute tubular injury, collapsing glomerulopathy, and extensive acute tubular injury through biopsy or autopsy. Besides, some studies have reported the presence of viral RNA or protein. However, some studies were unable to detect SARS-CoV-2 RNA in kidney tissue, urine, and serum.

Recent reports suggest that rhabdomyolysis and hyperkalemia occur in 7–20% and 23% of patients with evidence of COVID-19 AKI. The latter has been associated with metabolic acidosis. Biopsy samples on observation using light microscopy have demonstrated the evidence of injury that ranged from mild injury with ultrastructural evidence of glomerular subendothelial space widening and/or loss of endothelial cell fenestrae to severe injury manifest by endothelial cell swelling and fibrin thrombi within glomerular hilar arterioles and small arteries (Akilesh et al. 2021). In addition, postmortem findings for severe fatal COVID-19 revealed extensive ATI and a surprising endothelial injury pattern, with evidence for direct parenchymal tubular epithelial and podocyte viral infection on light microscopy (Su, Yang et al. 2020). In a study, Huang et al. (2021) performed ultrasonography for 390 patients six months after recovering from COVID-19. They reported no abnormality in the kidney morphology and size. Moreover, they reported that 13% without AKI and normal eGFR during the acute phase had decreased eGFR after six months. Decreased eGFR is a hallmark of AKI. Therefore, this necessitates follow-up of recovered patients to reduce morbidity and mortality.

8.2.5 Reproductive system

The SARS-CoV-2 infection has been shown to have various implications in different organs. The reproductive system, mainly the male reproductive system, is one of the most potential targets of SARS-CoV-2 due to the high abundance of TMPRSS2 and ACE2 receptor expression in testes (Wang and Xu 2020). There is still a conflict about the presence of SARS-COV-2 in semen samples. Diangeng et al. showed the presence of the virus in semen samples in 15.8% of a cohort from China by RT-PCR (Li et al. 2020). In contradiction to the previous study, Kayaslaan et al. (2020) confirmed the absence of the virus in semen samples collected from the Turkish population.

Despite the constant debate on viral protein detection in semen, the major concern in society is the assessment of virus-mediated male fertility. The pathological findings of SARS-CoV-2-infected patients' gonad signify the implications of the virus in male reproductive health. Yang et al. (2020) reported moderate to severe injury to Sertoli cells and seminiferous cells in 90% of the cohort included in the study with a significant reduction of Leydig cells. Histological signs of orchitis in testicles and microthrombi in prostates are two major observations made in infected patients (Duarte-Neto et al. 2020; Wichmann et al. 2020). Along with the histopathological study of the male gonads, assessing sex hormones would also provide an insight into their functionality. Patients with severe infection were shown to have low testosterone levels with high expression levels of luteinizing hormone, suggestive features of testes dysfunction (Madjunkov et al. 2020).

It would be interesting to investigate the negative impact of SARS-CoV-2 on semen parameters, including sperm count, motility, morphology, etc. Sperm count and sperm motility were shown to dwindle in patients with a moderate infection, while no such observations were made in patients with mild infection (Holtmann et al. 2020). Some studies investigated the impact of SARS-CoV-2 infection on female reproductive health. The expression level of viral receptor ACE2 is relatively lesser in the ovary, fallopian tube, vagina, and cervix compared to male gonads. A meta-analysis showed the prevalence of the infection in pregnant women constituting 10% of the total COVID-19-infected individuals (Allotey et al. 2020). The granular understanding of the negative impact of SARS-CoV-2 on the female reproductive system is yet to be evaluated; also the prolonged consequences of this viral infection are still unearthed.

8.2.6 Gastrointestinal tract

Diarrhea, nausea, anorexia, vomiting, or abdominal pain are some typical gastrointestinal (GI) tract symptoms manifested during COVID-19 infection. In a study evaluating 1,141 patients from Wuhan, China, 16% of the patients only presented gastrointestinal symptoms, and anorexia was the most common symptom (Luo et al. 2020). Diarrhea is the most common GI tract symptom associated with COVID-19, with a 5–10% pooled prevalence. Other typical symptoms such as anorexia, nausea, and vomiting, abdominal pain have a pooled prevalence of 26.8%, 7.85%, and 3.9–6.8%, respectively. Although all the studies present evidence for GI symptoms, none of them has documented the severity of these symptoms. Moreover, GI symptoms can develop for various reasons, and studies associating the cytopathic effects of the virus with the development of GI symptoms are limited. In addition, as COVID-19 is mainly associated with lower respiratory tract infection, patients presenting only gastrointestinal symptoms may be overlooked, delaying the diagnosis and further affecting the prognosis.

Viral entry into the host cells is crucial for the progression of the infection. The putative pathways for viral entry into the GI tract are via salivary secretions. However, the virus has to endure a highly acidic environment of the stomach. Darnell et al. (2004) observed viral inactivation at extreme acidic pH. Therefore, viral inactivation by gastric acid potentially inhibit viral entry, unless the patient has hypochlorhydria. However, in a study by Xiao et al. (2020), histological staining of esophageal, gastric, duodenal, and rectal tissues obtained using endoscopy showed the presence of ACE2 and viral nucleocapsid. ACE2 expression was detected in the upper esophagus, stratified epithelial cells, and absorptive enterocytes of the ileum and colon, duodenum, jejunum, and caecum (Dong et al. 2020). Lee et al. reported that the co-expressions of ACE2 and TMPRSS2 transcripts were mainly observed in the small intestine and colon using published scRNA-seq data and seven in-house normal colon samples. This indicated that the virus could infect the stomach tissues due to the high expression of ACE2 and TMPRSS2, which are required for viral entry. Altered gut flora due to increased release of pro-inflammatory cytokines and other markers of inflammation, use of antimicrobial medication for COVID-19, the gut–lung axis, and change in pathogenic organism ratio, aberrant mTORC activity due to ACE-2 receptor binding. However, these mechanisms are hypothetical and require reliable evidence to conclude their role in developing GI symptoms in COVID-19 patients. Recent article has summarized that viral RNA could be detected using fecal samples irrespective of the disease severity and even after the patient tested negative on nasopharyngeal swab RT-PCT test (Xiao et al. 2020). Moreover, the possibility of COVID-19 transmission through the fecal–oral route has been suggested by two studies (Trypsteen et al. 2020; Gavriatopoulou et al. 2020); however, substantial evidence is required for confirming this hypothesis. Therefore, patients showing GI tract symptoms should also be referred to the RT-PCR test for COVID-19 for early diagnosis.

8.2.7 Liver

The tropism of SARS-CoV-2 for liver tissue suggests that the liver can be a potential active proliferation site for viruses. The presence of SARS-CoV-2 RNA in stool and blood strengthens the possibility of direct viral infection in the liver (Yeo et al. 2020). An in-depth study on viral receptor expression in liver tissue suggested the enrichment of ACE2 receptors in cholangiocytes, thus providing an insight into the direct interaction of the SARS-CoV-2 virus with cholangiocytes to impair liver function (Chai et al. 2020). Lenti et al. (2020) performed a study on 100 consecutive patients to understand the impact of viral infection on liver function. They performed the routine biochemical test for the liver functional assessment, including the measurement of alanine aminotransferase (ALT), gamma-glutamyl transpeptidase, and total bilirubin. This study confirmed the alteration of liver function in 62.4% of patients at the stage of active infection which gradually improved after the recovery. As most of the studies reported the association of lung impairment with SARS-CoV-2 infection, the confirmation of viral clearance after their recovery is highly needed. An et al. (2021) compared the liver functional impairment during the hospitalization and after hospitalization at their recovery stage. The study showed very less improvement of liver function even after 40 days of post-COVID-19 infection. Nevertheless, the multiple factors responsible for the alteration of liver function, including epidemiological characteristics, and clinical manifestations, medications might worsen the disease outcomes. Thus, a comprehensive study investigating the long-term consequences of COVID-19 on liver function needs to be conducted.

8.3 Multi-omics strategy in COVID-19 research

The term "omics" implies the assessment of the global set of molecules present in an organism. Advancement of omics technologies, viz., genomics, transcriptomics, proteomics, and metabolomics, have enabled identifying disease biomarkers with high specificity and understanding the disease phenotype based on statistical inference. The combined analysis of single-nucleotide polymorphism (SNP), gene expression data, and RNA-seq data contribute to identifying the gene-level changes or genetic mutations in response to disease. In addition, Gene Ontology and KEGG pathway–based analysis provide an insight into the gene mutation–associated perturbed pathway. In this context, the terms "epigenomics" and "transcriptomics" shed light on the alteration of epigenetic modifications and transcript variants generated in response to complex disease. Several studies revealed the direct correlation of transcriptomics data with proteomics data. Herewith, the new omic technology "translatomics" has been brought forward which only focuses on the mRNA that translates into protein sequence. On the other hand, proteomics supplements other omic technologies to determine the structure, function, and modifications of the entire proteome present in a biospecimen and plays a crucial role in identifying disease-specific biomarkers. The alteration of protein functions also reflects at the metabolite level, hence the comprehensive metabolic analysis helps to identify the altered metabolic pathways in response to disease.

8.3.1 Genomics

Since the beginning of the COVID-19 pandemic, researchers all over the world are exploring multi-omics technologies to understand viral pathobiology. Among all the omics technologies, genomics has emerged as a promising strategy in deciphering the viral genome and its interaction with hosts. The high-throughput genomics technology, i.e., next-generation sequencing (NGS), helped researchers understand the viral genome's dynamicity and deducing the viral structural proteome. Different types of nucleic acid amplification tests (NAATs), including conventional RT-PCR, RT-LAMP, CRISPR technique, and Cartridge-based nucleic acid amplification test, are the most popular genomic techniques used for the diagnosis of viral infection with high specificity. Illumina COVID-Seq is a US-FDA approved NGS platform used to qualitatively detect the virus in respiratory specimens like- nasopharyngeal swabs. MinIon-based amplicon and metagenomic sequencing, nanopore target sequencing, and SARS-CoV-2 droplet digital PCR are modified genomics technologies developed for viral detection (Rai et al. 2021). A genomics-based study on SARS-CoV-2 shows the role of genome sequencing in determining the mutational degree of therapeutic targets, thus helping in the identification of potential therapeutic targets with less degree of mutation (Wang, Huang et al. 2020). Along with that, genomics technologies also assist in deducing the evolutionary trajectory and its epidemiological characteristics (Nakagawa and Miyazawa 2020). Wang, Hozumi et al. (2020) demonstrated the role of genomics technology in determining the phenotypic characteristics associated with disease severity. Using whole-genome sequencing, they identified *GOLGA3* and *DPP7* as two significant functional variants in COVID-19-infected patients. The comparative analysis between mild and severe patients showed a significant variation in the *IL-1* gene. No evidence on the investigation of long-term consequences of COVID-19 using advanced genomics technologies has been reported yet.

8.3.2 Transcriptomics

The transcriptome is dubbed as the snapshot in time of the total transcripts present in the cell. Microarrays and RNA-Seq are two fundamental contemporary transcriptomics techniques that use high-throughput sequencing to capture all sequences (Lowe et al. 2017). Transcriptomics analysis of virus-infected cells is essential for identifying host immune response dynamics and gene regulatory networks. Understanding the mechanism and host response against SARS-CoV-2 is necessary as the world still grapples with the pestilence (Islam et al. 2021). Intuitively, many researchers have performed transcriptomics analysis using various samples such as blood, peripheral blood mononuclear cells, nasopharyngeal swab, and bronchoalveolar lavage fluid. These studies have demonstrated that pro-inflammatory cytokine release might be a hallmark of COVID-19 patients and correlated with the disease severity. It was also observed that pathway alterations were more significant in severe patients, and a particular set of genes is induced to a greater extent by COVID-19. These expression signatures can be translated in a clinical setting for early prediction of disease outcomes and avoid fatal outcomes. Moreover, some researchers have also used transcriptomics for determining therapeutic targets or repurposable drugs using preexisting datasets (Jain et al. 2021).

In a single-cell RNA sequencing-based study, Pan et al. demonstrated that AKI in COVID-19 patients can be due to cytopathic effects of SARS-CoV-2 on podocytes and proximal straight tubule cells. Although their findings are essential for avoiding AKI in COVID-19 patients, the analysis was done using normal kidneys and evaluated the cytopathic effects of SARS-CoV-2 based on ACE2 and TMPRSS2 expression. Therefore, further studies demonstrating evidence of cytopathic effects of SARS-CoV-2 are necessary (Lowe et al. 2017). Studies assessing the extrapulmonary damage and long-term effects of COVID-19 are still lacking.

8.3.3 Proteomics

Proteomics has brought a paradigm shift in the way researchers think about biomarker discovery, and revolutionized the concept of precision medicine. Proteomics paves the way to combat COVID-19 by revealing the viral infection biology and host response to the infection. Lateral flow immunoassay is another advanced diagnostic assay developed for SARS-CoV-2 detection within 10–15 minutes (Pan et al. 2020). Advanced proteomics technologies like protein microarray and mass spectrometry have conquered the loopholes of conventional techniques in terms of protein quantification and characterization. Recently, Park et al. performed a global plasma proteomic analysis of COVID-19-infected patients to identify prognostic markers for disease severity. They reported dysregulation of the complement cascade, platelet function, neutrophil activation, and elevation of cytokine-like- interleukin 1, tumor necrosis factor in response to SARS-CoV-2 infection (Park et al. 2020). In addition, Cardozo showed the potentiality of mass spectrometry–based targeted proteomics for the detection of SARS-CoV-2 in respiratory samples with enduring specificity and sensitivity.

While most of the research has been focused on understanding the host response to viral infection, a few have been focused on the proteomics-based characterization of SARS-CoV-2-induced multi-organ injury. In a minireview, Lachen-Montes et al. articulated the correlation between olfactory loss and SARS-CoV-2 infection from a proteomics perspective (Lachen-Montes et al. 2020). A comprehensive study by Nie et al. (2021) revealed the alteration of protein profiling in multiple organs' autopsy samples of COVID-19-infected patients as compared to control. This study gives us an insight into the dysregulation of pathways in multiple organs, including glucose and fatty acid metabolism, hypoxia, coagulation, fibrosis, etc. Also, they reported the suppression of cholesterol biosynthesis in Leydig cells and the reduction of sperm motility. No such evidence on the persistent effects of the viral infection on multiple organs using proteomics technologies has been reported yet.

8.3.4 Metabolomics

Metabolomics refers to the comprehensive and quantitative analysis of all metabolites-small molecules (<1,000 Da) that play a crucial role in managing pathways in any organism in biological samples. Metabolites can be extracted from urine, saliva, or blood samples, which makes them easily accessible and desired biospecimens for biomarker search. Recent studies are mainly focused on understanding the effects of COVID-19 on metabolite profile of hospitalized individuals for studying the effect of disease severity and determining prognostic markers or therapeutic targets (Overmyer et al. 2020; Su, Chen et al. 2020; Sindelar et al. 2021; Roberts et al. 2020). Nonetheless, a recent study has reported the effect of COVID-19 patients three months after they were discharged from the hospital. The changes in the metabolome of the patients provided insights into the recovery from COVID-19. Even three months after discharge, the metabolite profile of recovered COVID-19 patients was different from a healthy control. On pathway analysis, metabolic alterations were observed in amino acids, purine, glycerophospholipid metabolism, citrate cycle (TCA cycle), and glyoxylate and dicarboxylate metabolism pathways. These alterations were closely related to pulmonary, hepatic, renal, microbial, energetic metabolism, and inflammation (Wang, Xu et al. 2020). Another study observed the metabolic profile of plasma from COVID-19 recovered patients three months after they were discharged. The cohort contained survivors who had pulmonary sequelae even three months after discharge. The alterations observed were mainly observed in amino acid and glycerophospholipid metabolic pathways; they were associated with the severity of the disease. The difference between the metabolic profile of survivors with pulmonary sequelae and healthy donors indicates that plasma metabolic profile can be used to monitor pulmonary dysfunction (Xu et al. 2021). Such studies can also be extended to understand the extrapulmonary effects of COVID-19, particularly in patients who have severe disease.

8.4 Future application of omics for the prognosis of disease consequences

The rapid and accurate investigation of post-COVID infection is crucial to combat the virus-mediated multi-organ injury. A few studies have been reported on the consequences of the diseases on vital organs like lungs, kidney, heart, and liver after one to three months of disease remission based on the clinical symptoms. Huang et al. first showed the consequences of viral infection six months postinfection. The persistence of muscle weakness, altered sleep cycle, and anxiety are the most common manifestations observed in COVID-19 survivors after six months of their recovery. The impaired lung function was the most crucial disease consequence that needed to be taken care of with high priority. The current approaches used for the determination of the long-term effect of COVID-19 in different organs include mostly CT-scan, X-ray imaging, and routine blood examination for different biochemical factors. The X-ray or CT of the chest is mostly used to determine the structural abnormalities of lungs within tissue like ground-glass opacities, pulmonary interstitial, and peripheral distributions that occurred due to COVID-19 infection. Total lung capacity, diffusion capacity, and forced expiratory volume are other parameters used for lung function tests. Electrocardiogram (ECG) and MRI are two extensively used tools that reflect the cardiac dynamicity in response to viral infection. Biochemical tests like ALT level, AST level, CRP level, troponin, albumin to creatinine ratio, and glomerular filtration rates are assessed to determine the functional changes in liver, heart, and kidney, respectively. The time consumption, laborious activities, availability of commercial kits, and specificity make the wide application of these techniques limited.

Recent studies have tried capturing the alterations in metabolic and proteomic profiles for COVID-19 survivors. These studies have addressed the fact that a robust body of evidence can aid in understanding the underlying mechanisms and determining prognostic markers. Therefore, multi-omics follow-up studies should be conducted in the future. These studies will help us decrease the effect of post-COVID-19 complications on the survivors' quality of life. The studies may include samples such as swabs, plasma, PBMCs, urine, or stools. In addition, in case of organ-specific complications, they can include biopsy samples (Figure 8.3). Currently, some studies have studied the extrapulmonary effects of COVID-19 using autopsy samples.

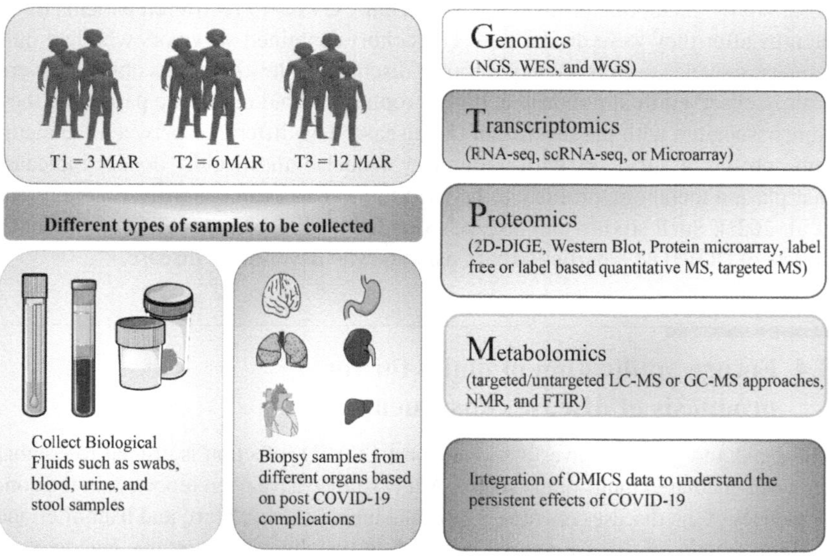

FIGURE 8.3 Putative multi-omics approach for understanding the post-COVID-19 complications. *Abbreviations:* MAR, months after recovery.

Source: This figure was created using Servier Medical Art templates, which are licensed under a Creative Commons Attribution 3.0 Unported License; https://smart.servier.com.

These studies have observed organ injury or dysregulation in pathways on histopathological and mass spectrometric analysis, respectively. These techniques can also be extrapolated to the biopsy samples to determine the dysregulated state's persistence at follow-ups. The availability of a sufficient biopsy sample to perform the multi-omics study is extremely challenging, hence investigating the alteration of circulatory biomarkers in follow-up cases is an unmet need. For example, Clara cell protein 6, surfactant protein D, desmosine, and matrix metalloproteinase 9 are established lung-specific markers. The levels of these proteins can be compared at different stages of infection and follow-up studies (Engström et al. 2012). In this regard, integrated omics-based investigation of COVID-19 survivors will provide a holistic overview of the long-term consequences of the viral infection and a way to combat the disease.

REFERENCES

Agricola, Eustachio, Alessandro Beneduce, Antonio Esposito, Giacomo Ingallina, Diego Palumbo, Anna Palmisano, Francesco Ancona et al. 2020. "Heart and Lung Multimodality Imaging in COVID-19." *JACC: Cardiovascular Imaging* 13 (8): 1792–808.

Akilesh, Shreeram, Cynthia C. Nast, Michifumi Yamashita, Kammi Henriksen, Vivek Charu, Megan L. Troxell, Neeraja Kambham et al. 2021. "Multicenter Clinicopathologic Correlation of Kidney Biopsies Performed in COVID-19 Patients Presenting with Acute Kidney Injury or Proteinuria." *American Journal of Kidney Diseases* 77 (1): 82–93.

Allotey, John, Elena Stallings, Mercedes Bonet, Magnus Yap, Shaunak Chatterjee, Tania Kew, Luke Debenham, et al. 2020. "Clinical Manifestations, Risk Factors, and Maternal and Perinatal Outcomes of Coronavirus Disease 2019 in Pregnancy: Living Systematic Review and Meta-Analysis." *BMJ* 370 (September): m3320.

An, Ya-Wen, Shuo Song, Wei-Xin Li, Yong-Xin Chen, Xiao-Peng Hu, Jia Zhao, Zhi-Wen Li, et al. 2021. "Liver Function Recovery of COVID-19 Patients after Discharge, a Follow-up Study." *International Journal of Medical Sciences* 18 (1): 176–86.

Arentz, Matt, Eric Yim, Lindy Klaff, Sharukh Lokhandwala, Francis X. Riedo, Maria Chong, and Melissa Lee. 2020. "Characteristics and Outcomes of 21 Critically Ill Patients With COVID-19 in Washington State." *JAMA: The Journal of the American Medical Association* 323 (16): 1612–14.

Argenziano, Michael G., Samuel L. Bruce, Cody L. Slater, Jonathan R. Tiao, Matthew R. Baldwin, R. Graham Barr, Bernard P. Chang et al. 2020. "Characterization and Clinical Course of 1000 Patients with Coronavirus Disease 2019 in New York: Retrospective Case Series." *BMJ* 369.

Balachandar, Vellingiri, Iyer Mahalaxmi, Subramaniam Mohana Devi, Jayaramayya Kaavya, Nachimuthu Senthil Kumar, Gracy Laldinmawii, Narayanasamy Arul et al. 2020. "Follow-up Studies in COVID-19 Recovered Patients—is It Mandatory?" *Science of the Total Environment*: 139021.

Bansal, Manish. 2020. "Cardiovascular Disease and COVID-19." *Diabetes & Metabolic Syndrome: Clinical Research & Reviews* 14 (3): 247–50.

Beyerstedt, Stephany, Expedito Barbosa Casaro, and Érika Bevilaqua Rangel. 2021. "COVID-19: Angiotensin-Converting Enzyme 2 (ACE2) Expression and Tissue Susceptibility to SARS-CoV-2 Infection." *European Journal of Clinical*

Microbiology & Infectious Diseases: Official Publication of the European Society of Clinical Microbiology 40 (5): 905–19. https://doi.org/10.1007/s10096-020-04138-6.

Bost, Pierre, Amir Giladi, Yang Liu, Yanis Bendjelal, Gang Xu, Eyal David, Ronnie Blecher-Gonen, et al. 2020. "Host-Viral Infection Maps Reveal Signatures of Severe COVID-19 Patients." *Cell* 181 (7): 1475–88.e12.

Bradley, Benjamin T., Heather Maioli, Robert Johnston, Irfan Chaudhry, Susan L. Fink, Haodong Xu, Behzad Najafian et al. 2020. "Histopathology and Ultrastructural Findings of Fatal COVID-19 Infections in Washington State: A Case Series." *The Lancet* 396 (10247): 320–32.

Chai, X., L. Hu, Y. Zhang, W. Han, Z. Lu, A. Ke, and J. Zhou. 2020. "Specific ACE2 Expression in Cholangiocytes May Cause Liver Damage after 2019-nCoV Infection." *Biorxiv*. www.biorxiv.org/content/10.1101/2020.02.03.931766v1.abstract.

Chen, Tao, Di Wu, Huilong Chen, Weiming Yan, Danlei Yang, Guang Chen, Ke Ma, et al. 2020. "Clinical Characteristics of 113 Deceased Patients with Coronavirus Disease 2019: Retrospective Study." *BMJ* 368 (March): m1091.

Cheng, Yichun, Ran Luo, Kun Wang, Meng Zhang, Zhixiang Wang, Lei Dong, Junhua Li, Ying Yao, Shuwang Ge, and Gang Xu. 2020. "Kidney Disease is Associated with In-hospital Death of Patients with COVID-19." *Kidney International* 97 (5): 829–38.

Darnell, Miriam E. R., Kanta Subbarao, Stephen M. Feinstone, and Deborah R. Taylor. 2004. "Inactivation of the Coronavirus That Induces Severe Acute Respiratory Syndrome, SARS-CoV." *Journal of Virological Methods* 121 (1): 85–91.

Dong, Mengzhen, Jie Zhang, Xuefeng Ma, Jie Tan, Lizhen Chen, Shousheng Liu, Yongning Xin, and Likun Zhuang. 2020. "ACE2, TMPRSS2 Distribution and Extra-pulmonary Organ Injury in Patients with COVID-19." *Biomedicine & Pharmacotherapy*: 110678.

Duarte-Neto, Amaro N., Renata A. A. Monteiro, Luiz F. F. da Silva, Denise M. A. C. Malheiros, Ellen P. de Oliveira, Jair Theodoro-Filho, João R. R. Pinho, et al. 2020. "Pulmonary and Systemic Involvement in COVID-19 Patients Assessed with Ultrasound-Guided Minimally Invasive Autopsy." *Histopathology* 77 (2): 186–97.

Engström, Gunnar, Claes Lindberg, Maria Gerhardsson de Verdier, Ulf Nihlén, Martin Anderson, Magnus Svartengren, and Kristina Forsman-Semb. 2012. "Blood Biomarkers and Measures of Pulmonary Function—A Study from the Swedish Twin Registry." *Respiratory Medicine* 106 (9): 1250–57.

Erickson, Michelle A., and William A. Banks. 2018. "Neuroimmune Axes of the Blood—Brain Barriers and Blood—Brain Interfaces: Bases for Physiological Regulation, Disease States, and Pharmacological Interventions." *Pharmacological Reviews* 70 (2): 278–314.

Feng, Zhichao, Qizhi Yu, Shanhu Yao, Lei Luo, Wenming Zhou, Xiaowen Mao, Jennifer Li et al. 2020. "Early Prediction of Disease Progression in COVID-19 Pneumonia Patients with Chest CT and Clinical Characteristics." *Nature Communications* 11 (1): 1–9.

Frija-Masson, Justine, Marie-Pierre Debray, Marie Gilbert, François-Xavier Lescure, Florence Travert, Raphaël Borie, Antoine Khalil, Bruno Crestani, Marie-Pia d'Ortho, and Catherine Bancal. 2020. "Functional Characteristics of Patients with SARS-CoV-2 Pneumonia at 30 Days Post-infection." *European Respiratory Journal* 56 (2).

Gavriatopoulou, Maria, Eleni Korompoki, Despina Fotiou, Ioannis Ntanasis-Stathopoulos, Theodora Psaltopoulou, Efstathios Kastritis, Evangelos Terpos, and Meletios A. Dimopoulos. 2020. "Organ-Specific Manifestations of COVID-19 Infection." *Clinical and Experimental Medicine* 20 (4): 493–506.

Hoffmann, Markus, Hannah Kleine-Weber, Simon Schroeder, Nadine Krüger, Tanja Herrler, Sandra Erichsen, Tobias S. Schiergens, et al. 2020. "SARS-CoV-2 Cell Entry Depends on ACE2 and TMPRSS2 and Is Blocked by a Clinically Proven Protease Inhibitor." *Cell* 181 (2): 271–80.e8. https://doi.org/10.1016/j.cell.2020.02.052.

Holtmann, Nora, Philippos Edimiris, Marcel Andree, Cornelius Doehmen, Dunja Baston-Buest, Ortwin Adams, Jan-Steffen Kruessel, and Alexandra Petra Bielfeld. 2020. "Assessment of SARS-CoV-2 in Human Semen-a Cohort Study." *Fertility and Sterility* 114 (2): 233–38.

Huang, Chaolin, Lixue Huang, Yeming Wang, Xia Li, Lili Ren, Xiaoying Gu, Liang Kang et al. 2021. "6-Month Consequences of COVID-19 in Patients Discharged from Hospital: A Cohort Study." *The Lancet* 397 (10270): 220–32.

Iadecola, Costantino, Josef Anrather, and Hooman Kamel. 2020. "Effects of COVID-19 on the Nervous System." *Cell* 183 (1): 16–27.e1.

Islam, Abul Bashar Mir Md Khademul, Md Abdullah-Al-Kamran Khan, Rasel Ahmed, Md Sabbir Hossain, Shah Md Tamim Kabir, Md Shahidul Islam, and Amam Zonaed Siddiki. 2021. "Transcriptome of Nasopharyngeal Samples from COVID-19 Patients and a Comparative Analysis with Other SARS-CoV-2 Infection Models Reveal Disparate Host Responses against SARS-CoV-2." *Journal of Translational Medicine* 19 (1): 1–25.

Jain, Ruchi, Sathishkumar Ramaswamy, Divinlal Harilal, Mohammed Uddin, Tom Loney, Norbert Nowotny, Hanan Alsuwaidi, et al. 2021. "Host Transcriptomic Profiling of COVID-19 Patients with Mild, Moderate, and Severe Clinical Outcomes." *Computational and Structural Biotechnology Journal* 19: 153–60.

Kayaaslan, Bircan, Gulay Korukluoglu, Imran Hasanoglu, Ayse Kaya Kalem, Fatma Eser, Esragul Akinci, and Rahmet Guner. 2020. "Investigation of SARS-CoV-2 in Semen of Patients in the Acute Stage of COVID-19 Infection." *Urologia Internationalis* 104 (9–10): 678–83.

Lachén-Montes, Mercedes, Fernando J. Corrales, Joaquín Fernández-Irigoyen, and Enrique Santamaría. 2020. "Proteomics Insights Into the Molecular Basis of SARS-CoV-2 Infection: What We Can Learn From the Human Olfactory Axis." *Frontiers in Microbiology* 11 (September): 2101.

Lee, Jaewon J., Scott Kopetz, Eduardo Vilar, John Paul Shen, Ken Chen, and Anirban Maitra. 2020. "Relative Abundance of SARS-CoV-2 Entry Genes in the Enterocytes of the Lower Gastrointestinal Tract." *Genes* 11 (6): 645.

Lenti, Marco Vincenzo, Federica Borrelli de Andreis, Ivan Pellegrino, Catherine Klersy, Stefania Merli, Emanuela Miceli, Nicola Aronico, et al. 2020. "Impact of COVID-19 on Liver Function: Results from an Internal Medicine Unit in Northern Italy." *Internal and Emergency Medicine* 15 (8): 1399–407.

Li, Diangeng, Meiling Jin, Pengtao Bao, Weiguo Zhao, and Shixi Zhang. 2020. "Clinical Characteristics and Results of Semen Tests Among Men with Coronavirus Disease 2019." *JAMA Network Open* 3 (5): e208292.

Liu, Ya-Fei, Zhe Zhang, Xiao-Li Pan, Guo-Lan Xing, Ying Zhang, Zhang-Suo Liu, and Sheng-Hao Tu. 2020. "The Chronic Kidney Disease and Acute Kidney Injury Involvement in COVID-19 Pandemic: A Systematic Review and Meta-analysis." *PLoS ONE* 16 (1): e0244779.

Lowe, Rohan, Neil Shirley, Mark Bleackley, Stephen Dolan, and Thomas Shafee. 2017. "Transcriptomics Technologies." *PLoS Computational Biology* 13 (5): e1005457.

Luo, Shihua, Xiaochun Zhang, and Haibo Xu. 2020. "Don't Overlook Digestive Symptoms in Patients with 2019 Novel Coronavirus Disease (COVID-19)." *Clinical Gastroenterology and Hepatology: The Official Clinical Practice Journal of the American Gastroenterological Association* 18 (7): 1636–37.

Madjunkov, Mitko, Michal Dviri, and Clifford Librach. 2020. "A Comprehensive Review of the Impact of COVID-19 on Human Reproductive Biology, Assisted Reproduction Care and Pregnancy: A Canadian Perspective." *Journal of Ovarian Research* 13 (1): 140.

Mo, Xiaoneng, Wenhua Jian, Zhuquan Su, Mu Chen, Hui Peng, Ping Peng, Chunliang Lei, Ruchong Chen, Nanshan Zhong, and Shiyue Li. 2020. "Abnormal Pulmonary Function in COVID-19 Patients at Time of Hospital Discharge." *European Respiratory Journal* 55 (6).

Nadim, Mitra K., Lui G. Forni, Ravindra L. Mehta, Michael J. Connor, Kathleen D. Liu, Marlies Ostermann, Thomas Rimmelé et al. 2020. "COVID-19-associated Acute Kidney Injury: Consensus Report of the 25th Acute Disease Quality Initiative (ADQI) Workgroup." *Nature Reviews Nephrology*: 1–18.

Nakagawa, So, and Takayuki Miyazawa. 2020. "Genome Evolution of SARS-CoV-2 and Its Virological Characteristics." *Inflammation and Regeneration* 40 (1): 17.

Ngai, Jenny C., Fanny W. Ko, Susanna S. Ng, Kin-Wang TO, Mabel Tong, and David S. Hui. 2010. "The Long-term Impact of Severe Acute Respiratory Syndrome on Pulmonary Function, Exercise Capacity and Health Status." *Respirology* 15 (3): 543–50.

Nie, Xiu, Liujia Qian, Rui Sun, Bo Huang, Xiaochuan Dong, Qi Xiao, Qiushi Zhang, et al. 2021. "Multi-Organ Proteomic Landscape of COVID-19 Autopsies." *Cell* 184 (3): 775–91.e14.

Oudit, G. Y., Z. Kassiri, C. Jiang, P. P. Liu, S. M. Poutanen, J. M. Penninger, and J. Butany. 2009. "SARS-coronavirus Modulation of Myocardial ACE2 Expression and Inflammation in Patients with SARS." *European Journal of Clinical Investigation* 39 (7): 618–25.

Overmyer, Katherine A., Evgenia Shishkova, Ian J. Miller, Joseph Balnis, Matthew N. Bernstein, Trenton M. Peters-Clarke, Jesse G. Meyer, et al. 2020. "Large-Scale Multi-Omic Analysis of COVID-19 Severity." *medRxiv: The Preprint Server for Health Sciences*, July. https://doi.org/10.1101/2020.07.17.20156513.

Pan, Xiu-Wu, Da Xu, Hao Zhang, Wang Zhou, Lin-Hui Wang, and Xin-Gang Cui. 2020. "Identification of a Potential Mechanism of Acute Kidney Injury during the COVID-19 Outbreak: A Study Based on Single-Cell Transcriptome Analysis." *Intensive Care Medicine* 46 (6): 1114–16.

Park, Joonho, Hyeyoon Kim, So Yeon Kim, Yeonjae Kim, Jee-Soo Lee, Kisoon Dan, Moon-Woo Seong, and Dohyun Han. 2020. "In-Depth Blood Proteome Profiling Analysis Revealed Distinct Functional Characteristics of Plasma Proteins between Severe and Non-Severe COVID-19 Patients." *Scientific Reports* 10 (1): 22418.

Prasad, Ashish, and Manoj Prasad. 2020. "Single Virus Targeting Multiple Organs: What We Know and Where We Are Heading?" *Frontiers in Medicine* 7: 370.

Puntmann, Valentina O., M. Ludovica Carerj, Imke Wieters, Masia Fahim, Christophe Arendt, Jedrzej Hoffmann, Anastasia Shchendrygina et al. 2020. "Outcomes of Cardiovascular Magnetic Resonance Imaging in Patients Recently Recovered from Coronavirus Disease 2019 (COVID-19)." *JAMA Cardiology* 5 (11): 1265–73.

Rai, Praveen, Ballamoole Krishna Kumar, Vijaya Kumar Deekshit, Indrani Karunasagar, and Iddya Karunasagar. 2021. "Detection Technologies and Recent Developments in the Diagnosis of COVID-19 Infection." *Applied Microbiology and Biotechnology* 105 (2): 441–55.

Richardson, Safiya, Jamie S. Hirsch, Mangala Narasimhan, James M. Crawford, Thomas McGinn, Karina W. Davidson, Douglas P. Barnaby et al. 2020. "Presenting Characteristics, Comorbidities, and Outcomes Among 5700 Patients Hospitalized with COVID-19 in the New York City Area." *JAMA* 323 (20): 2052–59.

Roberts, Ivayla, Marina Wright Muelas, Joseph M. Taylor, Andrew S. Davison, Yun Xu, Justine M. Grixti, Nigel Gotts, Anatolii Sorokin, Royston Goodacre, and Douglas B. Kell. 2020. "Untargeted Metabolomics of COVID-19 Patient Serum Reveals Potential Prognostic Markers of Both Severity and Outcome." *medRxiv*. www.medrxiv.org/content/10.1101/2020.12.09.20246389v1.abstract.

Shaha, Kunal Bikram, Dhiraj Narayan Manandhar, Jung Rae Cho, Ashok Adhikari, and Man Bahadur KC. 2020. "COVID-19 and the Heart: What We Have Learnt So Far." *Postgraduate Medical Journal* 97 (1152): 655–66.

Sindelar, Miriam, Ethan Stancliffe, Michaela Schwaiger-Haber, Dhanalakshmi S. Anbukumar, Randy A. Albrecht, Wen-Chun Liu, Kayla Adkins Travis, Adolfo García-Sastre, Leah P. Shriver, and Gary J. Patti. n.d. "Longitudinal Metabolomics of Human Plasma Reveals Robust Prognostic Markers of COVID-19 Disease Severity." *medRxiv*. https://doi.org/10.1101/2021.02.05.21251173.

Su, Hua, Ming Yang, Cheng Wan, Li-Xia Yi, Fang Tang, Hong-Yan Zhu, Fan Yi et al. 2020. "Renal Histopathological Analysis of 26 Postmortem Findings of Patients with COVID-19 in China." *Kidney International* 98 (1): 219–27.

Su, Yapeng, Daniel Chen, Dan Yuan, Christopher Lausted, Jongchan Choi, Chengzhen L. Dai, Valentin Voillet, et al. 2020. "Multi-Omics Resolves a Sharp Disease-State Shift between Mild and Moderate COVID-19." *Cell* 183 (6): 1479–95.e20.

Suryawanshi, Hemant, Pavel Morozov, Thomas Tuschl, Thangamani Muthukumar, Masashi Yamaji, and Zev Williams. 2020. "Cell-type-specific Expression of Renin-angiotensin-system Components in the Human Body and Its Relevance to SARS-CoV-2 Infection." *BioRxiv*.

Torres-Castro, R., L. Vasconcello-Castillo, X. Alsina-Restoy, L. Solis-Navarro, F. Burgos, H. Puppo, and J. Vilaró. 2020. "Respiratory Function in Patients post-infection by COVID-19: A Systematic Review and Meta-analysis." *Pulmonology* 27 (4): 328–37.

Trypsteen, Wim, Jolien Van Cleemput, Willem van Snippenberg, Sarah Gerlo, and Linos Vandekerckhove. 2020. "On the Whereabouts of SARS-CoV-2 in the Human Body: A Systematic Review." *PLoS Pathogens* 16 (10): e1009037.

Wang, Dawei, Bo Hu, Chang Hu, Fangfang Zhu, Xing Liu, Jing Zhang, Binbin Wang et al. 2020. "Clinical Characteristics of 138 Hospitalized Patients with 2019 Novel Coronavirus—Infected Pneumonia in Wuhan, China." *JAMA* 323 (11): 1061–69.

Wang, Fang, Shujia Huang, Rongsui Gao, Yuwen Zhou, Changxiang Lai, Zhichao Li, Wenjie Xian, et al. 2020. "Initial Whole-Genome Sequencing and Analysis of the Host Genetic Contribution to COVID-19 Severity and Susceptibility." *Cell Discovery* 6 (1): 83.

Wang, Rui, Yuta Hozumi, Changchuan Yin, and Guo-Wei Wei. 2020. "Mutations on COVID-19 Diagnostic Targets." *Genomics* 112 (6): 5204–13.

Wang, Sufei, Juanjuan Xu, Ping Luo, Lian Yang, Siwei Song, Xueyun Tan, Qing Chen, et al. 2020. "Plasma Metabolomic Profiles and Clinical Features in Recovered COVID-19 Patients Without Previous Underlying Diseases 3 Months After Discharge." www.researchsquare.com/article/rs-110981/latest.pdf.

Wang, Zhengpin, and Xiaojiang Xu. 2020. "scRNA-Seq Profiling of Human Testes Reveals the Presence of the ACE2 Receptor, A Target for SARS-CoV-2 Infection in Spermatogonia, Leydig and Sertoli Cells." *Cells* 9 (4). https://doi.org/10.3390/cells9040920.

Wichmann, Dominic, Jan-Peter Sperhake, Marc Lütgehetmann, Stefan Steurer, Carolin Edler, Axel Heinemann, Fabian Heinrich, et al. 2020. "Autopsy Findings and Venous Thromboembolism in Patients With COVID-19: A Prospective Cohort Study." *Annals of Internal Medicine* 173 (4): 268–77.

Wiersinga, W. J., A. Rhodes, A. C. Cheng, S. J. Peacock, and H. C. Prescott. 2020. "Pathophysiology, Transmission, Diagnosis, and Treatment of Coronavirus Disease 2019 (COVID-19): A Review." *JAMA* 324 (8): 782–93. https://doi.org/10.1001/jama.2020.12839.

Xiao, Fei, Meiwen Tang, Xiaobin Zheng, Ye Liu, Xiaofeng Li, and Hong Shan. 2020. "Evidence for Gastrointestinal Infection of SARS-CoV-2." *Gastroenterology* 158 (6): 1831–33.

Xu, Juanjuan, Mei Zhou, Ping Luo, Zhengrong Yin, Sufei Wang, Tingting Liao, Fan Yang, et al. 2021. "Plasma Metabolomic Profiling of Patients Recovered from COVID-19 with Pulmonary Sequelae 3 Months after Discharge." *Clinical Infectious Diseases* 73 (12): 2228–39. https://doi.org/10.1093/cid/ciab147.

Xu, Zhe, Lei Shi, Yijin Wang, Jiyuan Zhang, Lei Huang, Chao Zhang, Shuhong Liu et al. "Pathological Findings of COVID-19 Associated with Acute Respiratory Distress Syndrome." *The Lancet Respiratory Medicine* 8 (4): 420–22.

Yang, Ming, Shuo Chen, Bo Huang, Jing-Min Zhong, Hua Su, Ya-Jun Chen, Qin Cao et al. 2020. "Pathological Findings in the Testes of COVID-19 Patients: Clinical Implications." *European Urology Focus* 6 (5): 1124–29.

Yeo, Charleen, Sanghvi Kaushal, and Danson Yeo. 2020. "Enteric Involvement of Coronaviruses: Is Faecal-Oral Transmission of SARS-CoV-2 Possible?" *The Lancet. Gastroenterology & Hepatology* 5 (4): 335–37.

Zhang, Yuhao, Xiuchao Geng, Yanli Tan, Qiang Li, Can Xu, Jianglong Xu, Liangchao Hao et al. 2020. "New Understanding of the Damage of SARS-CoV-2 Infection Outside the Respiratory System." *Biomedicine & Pharmacotherapy*: 110195.

Zhao, Yu-miao, Yao-min Shang, Wen-bin Song, Qing-quan Li, Hua Xie, Qin-fu Xu, Jun-li Jia et al. "Follow-up Study of the Pulmonary Function and Related Physiological Characteristics of COVID-19 Survivors Three Months After Recovery." *EClinicalMedicine* 25 (2020): 100463.

Zou, Xin, Ke Chen, Jiawei Zou, Peiyi Han, Jie Hao, and Zeguang Han. 2020. "Single-cell RNA-Seq Data Analysis on the Receptor ACE2 Expression Reveals the Potential Risk of Different Human Organs Vulnerable to 2019-nCoV Infection." *Frontiers of Medicine*: 1–8.

9

Insights into interactomics-driven drug repurposing to combat COVID-19

Amrita Mukherjee, Ayushi Verma, Ananya Burli, Krishi Mantri, and Surbhi Bihani

CONTENTS

DOI: 10.1201/9781003220787-9

List of abbreviations

BE – Binding energy
DTI – Drug target information
PLIP – Protein interaction ligand profiler
SARS-CoV-2 – Severe acute respiratory syndrome coronavirus 2

9.1 Proteins, diseases, and drug action

The great majority of pharmacologically active compounds currently in use bind and modulate the activity of proteins. The recent proteomic technologies allow for the holistic characterization of many proteins and have proven to be valuable in practically every aspect of the drug discovery and development program. Further, the protein–protein interactions are crucial because many proteins function by forming biocomplexes (Kuzmanov and Emili 2013). Altered protein activity, be it in terms of function, structure, or protein–protein interaction, is at the core of any medical condition (Frantzi, Latosinska, and Mischak 2019). The four types of macromolecules (proteins, nucleic acids, polysaccharides, and lipids) that can be targeted for therapeutic intervention, proteins represent the largest class of targets amenable to pharmacological modulation (Hopkins and Groom 2002).

9.2 Omics-based approaches for COVID-19 drug repurposing

9.2.1 Role of clinical proteomics in targeted drug repurposing

Proteomic studies have led to the identification of altered host pathways, including but not limited to neutrophil activation and degranulation, complement system, blood coagulation, and hemostasis, lipid metabolism, macrophage function, platelet aggregation, and degranulation, amino acid metabolism, inflammation, and acute-phase response, cytokine and interferon signaling, and T-cell-mediated immunity, the dysregulation of which underlie the pathophysiology of COVID-19 (Barh et al. 2020; D'Alessandro et al. 2020; Leng et al. 2020; Messner et al. 2020; Filbin et al. 2020; Shen et al. 2020; Su et al. 2020; Hou et al. 2020; Chen et al. 2020; Li, Zhao, and Zhan 2021). The drugs targeting these pathways are an effective strategy to reduce the severity of disease and combat infection. For instance, FDA-approved anti-inflammatory drugs like Colchicine and Anakinra are undergoing clinical trials to treat COVID-19 (ClinicalTrials.gov identifiers: NCT04326790 and NCT04603742, respectively). Dipyridamole is another FDA-approved drug and an inhibitor of blood coagulation that has shown marked clinical improvement in severely ill COVID-19 patients (X. Liu et al. 2020). Elevated interleukin 6 and 8 (IL-6 and IL-8) levels, which are responsible for excessive infiltration of neutrophils at the site of infection, were identified to be associated with severe COVID-19 (Li et al. 2021; Del Valle et al. 2020). Tocilizumab, an anti-IL-6 monoclonal antibody, and an anti-IL-8 antibody are under consideration to control the inflammatory response in COVID-19 patients (ClinicalTrials.gov Identifier: NCT04317092 and NCT04347226, respectively) (Stone et al. 2020; Luo et al. 2020). Thus, clinical proteomics can prove highly advantageous for the speedy identification of therapeutic targets for drug repurposing.

9.2.2 *In vitro* SARS-CoV-2-infected cell culture model for proteomics analysis

Cell line–based *in vitro* infection models require minimal ethical clearance and can be set up quickly. Therefore, proteome analysis of *in vitro* infection models has been utilized as a rapid means to conduct preliminary studies to identify fundamental cellular processes and host factors affected during infection. For instance, Bojkova et al. looked at the perturbations in the proteome of Caco-2 cells, a human cancer cell line, in response to SARS-CoV-2 infection, revealing critical cellular pathways, viz., carbon metabolism, nucleotide biosynthesis, splicing, and proteostasis, among others, pertinent to viral pathogenesis (Bojkova et al. 2020). They studied the antiviral activity of inhibiting these pathways on SARS-CoV-2 infection, thereby identifying new drugs and drug targets (Bojkova et al. 2020). They observed inhibition of SARS-CoV-2 by Ribavirin (an FDA-approved antiviral drug and an inhibitor of nucleotide synthesis) and 2-deoxy-d-glucose (an anticancer drug and an inhibitor of carbon metabolism), both of which are undergoing clinical trials for COVID-19 (NCT04494399, CTRI/2020/06/025664) (Hung et al. 2020; Verma et al. 2020). In another study, Bouhaddou et al. investigated the phosphoproteome of SARS-CoV-2 infected Vero E6 cells (a cell line originating from an African green monkey) using Mass Spectrometry (MS)–based approach (Bouhaddou et al. 2020). They studied alterations in the phosphorylation pattern of proteins and mapped the changes to disrupted kinases and pathways, highlighting how the virus utilizes the posttranslational regulatory systems to orchestrate changes in cellular signaling (Bouhaddou et al. 2020). They further identified 68 inhibitors against the most differentially regulated kinases and tested them for cellular toxicity and antiviral activity resulting in the prioritization of several FDA-approved drugs or at different stages of clinical development (Bouhaddou et al. 2020). Thus, proteomics of infection models is an attractive strategy to study the cellular processes modulated in response to infection and identify therapeutic targets for drug repurposing.

9.2.3 Interactomics: Discipline connecting biology with informatics

Technological advances in omics-based technologies have enabled the characterization of the interactome, leading to a field of study called "interactomics", which has been instrumental in resolving the functional diversity and complexity of a cell at a whole new level. Some of the biological processes indispensable to the viral life cycle are performed by protein complexes of multiple units rather than single proteins. For example, non-structural proteins 7 and 8 (nsp7 and nsp8) aid in the RNA-dependent RNA polymerase function of nsp12 by forming an essential protein complex (Gao et al. 2020). Disruption of cardinal virus–virus or host–virus interactions is an effective strategy to combat infection. Interactomics that seek to characterize such interactions has been greatly facilitated by affinity–purification mass spectrometry. Gordon et al. expressed affinity-tagged SARS-CoV-2 proteins in human cells and co-purified human proteins interacting with the viral proteins by affinity chromatography, and then identifying the human proteins by mass spectrometry (Gordon et al. 2020). This led to a host–virus protein–protein interaction network that has been instrumental in elucidating the molecular mechanism of infection and identifying targets for drug repurposing. The study identified 69 compounds, including 29 FDA-approved drugs, 12 clinical, and 28 preclinical compounds targeting the host proteins and screened a group of them for their antiviral activity. In another approach, Schmidt et al. utilized RNA antisense purification and mass spectrometry to purify and characterize host cellular proteins directly binding to SARS-CoV-2 RNAs (Schmidt et al. 2021). They further demonstrated pharmacological inhibition of SARS-CoV-2 replication by targeting some of the RNA-binding host proteins, thereby providing viable therapeutic targets for drug repurposing (Schmidt et al. 2021).

9.2.4 Drug signature mapping

Proteomics-based approaches (Figure 9.1) allow for an unbiased, data-driven target identification that would eventually raise the success rate of the drug repositioning program. Recently, scientists were able to identify Ivermectin, a broad-spectrum, FDA-approved antiparasitic drug, as a possible treatment option for COVID-19 using repurposing approach (Li, Zhao, and Zhan 2021). Using label-based quantitative proteomics, they studied the proteome signature in response to Ivermectin and identified many proteins dysregulated due to SARS-CoV-2 infection (Bojkova et al. 2020) to be involved in Ivermectin-regulated pathways. In another study, Ivermectin was shown to inhibit SARS-CoV-2 replication *in vitro* (Caly et al. 2020) and was found to be safe and effective for treating mild COVID-19 in a small group of patients in Bangladesh (Ahmed et al. 2021). The clinical trial results for ivermectin are currently awaited (ClinicalTrials.gov Identifier: NCT04668469).

9.3 *In silico*/computational approaches: Drug repurposing

Data from proteomics and protein network studies are rapidly accumulating and are being made accessible. The general workflow includes three significant

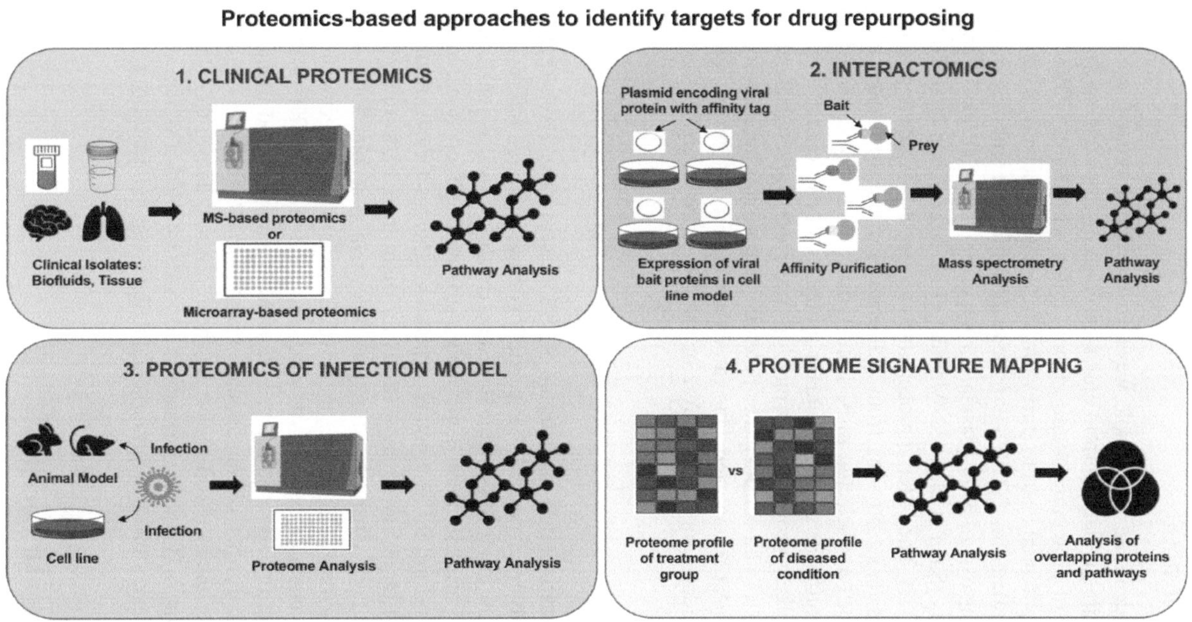

FIGURE 9.1 Interactomics-based approaches to identify targets for drug repurposing. The various proteomics-based approaches to identify the target proteins for drug repurposing: 1. Clinical proteomics involves the sample preparation from clinical isolates and proteomic analysis to decipher the target proteins. 2. Interactomics involves the expression of tagged bait protein (of pathogen) in the cell line followed by co-elution of interacting prey proteins (of host) by affinity purification. The approach is frequently used to identify host–pathogen interaction. 3. In another approach, the cell line or the animal model is infected with the pathogen and used for proteomics analysis. 4. Proteome signature mapping involves the comparison of proteome signatures of diseased groups and drug treatment groups for the identification of common proteins.

steps: drug-target identification, drug repurposing, and side effect assessment (Hodos et al. 2016). In this chapter, we will discuss the first two steps in detail.

9.3.1 Target identification

Drug targets can be identified in multiple ways. In the first approach, one can select target proteins directly derived from omics-based studies (proteomics, interactomics, etc.). Plenty of proteomics studies from clinical samples of COVID-19 patients have given us potential leads to start with. They majorly include the human proteins that are upregulated in COVID-19 patients postinfection and they differentiate a biological sample (e.g., swab, plasma, serum, urine) from COVID-19 positive to negative patients (Shen et al. 2020; Messner et al. 2020; Chen et al. 2020; Li, Zhao, and Zhan 2021). The latter approach includes intense literature mining, where the well-studied target proteins from a disease or disease-causing pathogens (e.g., SARS-CoV-2) are selected to be inhibited by small molecules. Mpro (3-chymotrypsin like protease or 3CLpro/main protease), RNA-dependent RNA polymerase (RdRp), spike glycoprotein (S), envelope protein (E), and nucleocapsid (N) from SARS-CoV-2 can be considered as potential drug targets here. Choosing the set of drugs is the next crucial step. One can either make a customized library as probable drug candidates for SARS-CoV-2 or use the entire database of small molecules.

9.3.2 Databases

Three-dimensional structures are available in RCSB Protein Data Bank (PDB), which are solved mainly by X-ray or NMR crystallography. Similarly, information about small molecules or drugs is available in diverse databases such as PubChem, Drugbank, ZINC, and ChEMBL. PubChem contains structural and functional information for more than 60 million compound structures, whereas Drugbank offers multiple levels of drug target information (DTI) for more than 10,000 drugs. In PubChem, we can find both 3D and 2D structures of a small molecule. Sometimes instead of 3D structures (e.g., SDF format), the SMILES (simplified molecular-input line-entry system) are available, which have to be converted to either SDF or PDB format. All these structures can be retrieved from these databases and manually curated before starting further computational experiments. Recently, a customized SARS-CoV-2 drug interaction database has been created, named CORDITE (Curated CORona Drug InTERactions Database for SARS-CoV-2) (Martin et al. 2020). This platform is very much helpful in conducting a curated literature search for COVID-19.

9.3.3 *In silico* molecular docking

Once the target proteins and ligands are shortlisted, *in silico* molecular docking is performed. This is a method by which the molecular recognition process is computationally simulated (Figure 9.2). During this procedure, several optimized conformations between protein and its ligands are optimized. Such

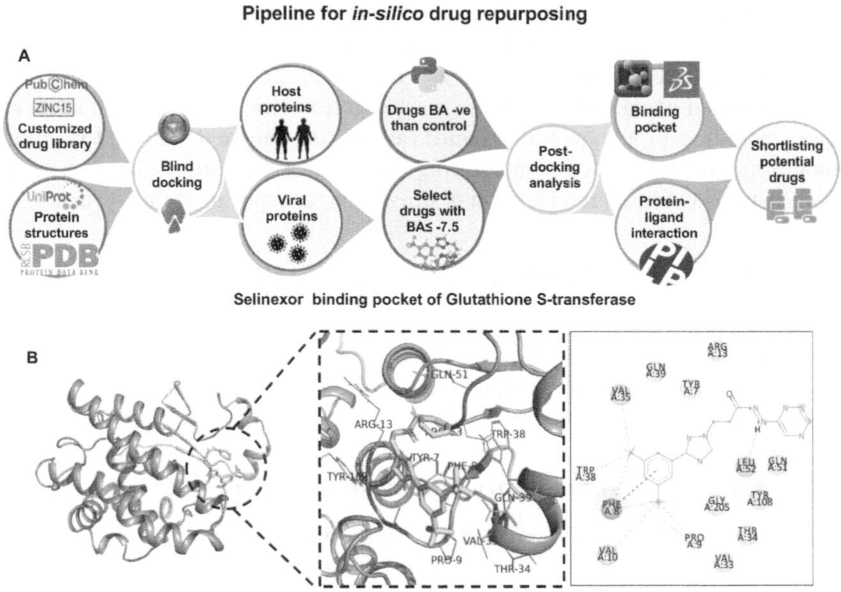

FIGURE 9.2 ***In silico*** molecular docking (pipeline and results). (a) Pipeline for *in silico* drug repurposing by molecular docking. (b) Docking results showing 3D and 2D representation of protein–ligand interaction of Selinexor (CID 71481097) bound with target protein (chosen from host) glutathione S-transferase P at the ligand-binding pocket.

optimization generates multiple "protein-ligand docked poses", each having a score, generally named as "binding energy". The lower the score, the better the binding energy is considered (Aamir et al. 2018). One can use a cutoff value for the binding energy score as per the study type and requirements. Sometimes, information about the ligand-binding site of a particular protein is quite helpful. It is required to improve the sampling efficiency of the docking program (Kong et al. 2019). In that case, site-specific docking is being performed. When enough information about ligand binding pockets is not available, one can go for "blind docking". In this approach, the entire length of the protein is included in the docking grid box. The software usually used to perform docking is described in Section 3.3.

9.4 Drug repurposing in the light of COVID-19: Methodologies

9.4.1 MS-based proteomics for target identification

In the middle of the year 2020, an extensive proteomics-based investigation of nasopharyngeal swab and plasma samples of COVID-19 patients was performed at Proteomics Laboratory, IIT Bombay (Bankar et al. 2021; Suvarna et al. 2021). Proteins were extracted from virus-inactivated clinical samples by three different protein extraction strategies (Xu and Hagler 2002). Extracted proteins were then identified using high-resolution mass spectrometric methods such as shotgun (label-free protein/peptide quantification), and targeted (selected reaction monitoring/multiple reaction monitoring) proteomics. This study gave us a comprehensive list of host proteins from *in silico* drug repurposing studies (Table 9.1).

9.4.2 Selection of SARS-CoV-2 proteins from literature

A total of ten viral proteins were selected from SARS-CoV-2 protein database, including the main protease (Nsp5), RNA-dependent RNA polymerase, envelope, nucleocapsid, spike glycoprotein, and few other non-structural proteins, essential for viral replication (Table 9.2).

9.4.3 Curation of drug library and protein structures

Before proceeding with *in silico* drug repurposing using a molecular docking strategy, it is necessary to create a drug repository containing those drugs needed to be screened against the differentially expressed protein targets found from the proteomics analysis in the previous sections (Figure 9.2a). To make a customized

TABLE 9.1

List of Viral Proteins Used in *In Silico* Molecular Docking

S. No.	ORF Name	Protein Name	Uniport ID	Crystal Structure	Control Inhibitors
1	Nsp1	Replicase polyprotein 1ab	P0C6X7	6ZLW(i) 1–180	Tirilazad (DB13050)
2	Nsp3 (Macro X domain)	Replicase polyprotein 1ab	P0DTD1	6WEY	CQN2H (C19H15ClN2O2)
3	Nsp5	Main protease	P0DTD1	7BUY	Carmofur
4	Nsp9	3c-like proteinase peptide, non-structural protein 9 fusion	P0DTD1	6W9Q, PDB ID-6W4B)	Conivaptan
5	Nsp10		P0DTD1	6W4H, PDB ID: 6W75	Eriodyctiol
6	Nsp12	RNA-dependent RNA polymerase	P0DTD1	7BV2	Nilotinib
7	Nsp13	Helicase protein	P0DTD1	6XEZ	Cepharanthine
8	Nsp16		P0DTD1	6W4H, 6W75A	Eriodyctiol
9	N	Nucleocapsid	P0DTC9	6VYO	Zidovudine
10	E	Envelope small-membrane protein	P0DTC4	7K3G	Hexamethylene amiloride (HMA)

library, deep literature mining was done. A total of 58 drugs (Table 9.3) were selected from a recent study by Gordon et al., where the chemoinformatic analysis was performed with these drugs against 332 high-confidence protein–protein interactions between SARS-CoV-2 and human proteins (Gordon et al. 2020).

9.4.4 *In silico* drug repurposing strategies using molecular docking software

9.4.4.1 PyRx

PyRx is an integrated virtual screening software used in computational drug discovery. It is a convenient application supplemented with a powerful visualization engine and spreadsheet-like functionality that enables users to effectively analyze

TABLE 9.2

List of Host Proteins Used in *In Silico* Molecular Docking

Proteins	UniProt ID	Control Inhibitor	Predicted Inhibitors
HSP 90-alpha	P07900	Geldanamycin	Selinexor, dabrafenib, ponatinib, silmitasertib
Calcyphosin	Q13938	Paroxetine	Selinexor, dabrafenib, ponatinib, silmitasertib, daunorubicin
Carbonic anhydrase 2	P00918	Aspirin, acetazolamide	Selinexor, dabrafenib, silmitasertib
Glutathione S-transferase P	P09211	Ethacrynic acid	Selinexor, dabrafenib, ponatinib, silmitasertib, daunorubicin
Peroxiredoxin-1	P32119	β-Elemene	Dabrafenib, daunorubicin
Pyruvate kinase PKM	P14618	l-Phenylalanine	None
Triosephosphate isomerase	P60174	PEP	Daunorubicin
Aldehyde dehydrogenase	P30838	Disulfiram	Selinexor, dabrafenib, ponatinib, daunorubicin
Transketolase	P29401	Oxythiamine	Dabrafenib, ponatinib, daunorubicin
l-Lactate dehydrogenase A chain	P00338	FX11	Selinexor, dabrafenib, ponatinib, silmitasertib, daunorubicin
Clathrin heavy chain 1	Q00610	ES9–17	Ponatinib, silmitasertib, daunorubicin
Fascin	Q16658	NP-G2–044	None
HSP 90-beta	P08238	Geldanamycin	None

TABLE 9.3
List of Drugs Used in *In Silico* Molecular Docking

S. No.	Drugs	ZINC ID	Status
1	Chloroquine	ZINC19144231	Approved
2	Dabrafenib	ZINC68153186	Approved
3	Captopril	ZINC57001	Approved
4	Lisinopril	ZINC3812863	Approved
5	Camostat	ZINC3871842	Approved
6	Nafamostat	ZINC3874467	Approved
7	Chloramphenicol	ZINC113382	Approved
8	Tigecycline	ZINC14879972	Approved
9	Linezolid	ZINC2008866	Approved
10	Minoxidil	ZINC1735	Approved
11	Selinexor	ZINC96170454	Approved
12	Silmitasertib	ZINC58638454	Approved
13	Valproic Acid	ZINC3008621	Approved
14	Haloperidol	ZINC537822	Approved
15	Loratadine	ZINC537931	Approved
16	Entacapone	ZINC35342787	Approved
17	Indomethacin	ZINC601283	Approved
18	Metformin	ZINC12859773	Approved
19	Ponatinib	ZINC000036701290	Approved
20	Migalastat	ZINC1636704	Approved
21	Mycophenolic acid	ZINC000000001758	Approved
22	Ribavirin	ZINC1035331	Approved
23	Azacitidine	ZINC3861768	Approved
24	Ruxolitinib	ZINC43207851	Approved
25	Daunorubicin	ZINC3917708	Approved
26	S-Verapamil	ZINC3812888	Approved
27	EGCG	ZINC3870412	Approved
28	Quercetin	ZINC3869685	Approved
29	Rapamycin	ZINC169289388	Approved
30	Anisomysin	ZINC954	Approved
31	Sapanisertib	ZINC73069271	Clinical
32	Pevonedistat	ZINC58660702	Clinical
33	RVX-208	ZINC43199551	Clinical
34	E-52862	ZINC95000617	Clinical
35	Merimepodib	ZINC3975663	Clinical
36	ABBV-744	ZINC1250228389	Clinical
37	CPI-0610	ZINC200480149	Clinical
38	CB5083	ZINC208076131	Clinical
39	Tomivosertib	ZINC575623807	Clinical
40	Bafilomycin A1	ZINC000169647947	Preclinical
41	CCT 365623	ZINC1776011385	Preclinical

S. No.	Drugs	ZINC ID	Status
42	AZ8838	ZINC914431265	Preclinical
43	AZ3451	ZINC914431341	Preclinical
44	4E2RCat	ZINC7018722	Preclinical
45	DBeQ	ZINC39389	Preclinical
46	ML240	ZINC96021026	Preclinical
47	JQ1	ZINC57318556	Preclinical
48	TMCB	ZINC000058638742	Preclinical
49	Apicidin	ZINC17654900	Preclinical
50	RS-PPCC	ZINC28635556	Preclinical
51	PB28	ZINC22931094	Preclinical
52	H-89	ZINC2043204	Preclinical
53	ZINC4326719	ZINC4326719	Preclinical
54	ZINC4511851	ZINC4511851	Preclinical
55	ZINC95559591	ZINC95559591	Preclinical
56	AC-55541	ZINC13302309	Preclinical
57	GB110	ZINC64527010	Preclinical
58	UCPH-101	ZINC000040914195	Preclinical

the drug molecules' binding to the corresponding proteins. PyRx comes with multiple features inculcated into it in the form of open-source software that aids in the procedure. These include:

- *AutodockVina:* Software used for docking using the Vina Wizard
- *OpenBabel:* For importing .sdf files, minimizing energy
- *Visualization ToolKit (VTK):* For visualization of docked structures, generating 3D images post-docking.

The docking of the drug library generates nine different conformers of each drug, known as "poses". These poses, numbered from 0–8, each have a different binding affinity to the protein, even if they describe the binding of the same drug. The post-docking file generated for each drug comprises all nine poses. Pose "0" for a drug possesses the information for the conformer with the most negative binding energy and highest binding affinity and was therefore selected to compare the binding affinities of various drugs in the library to that of the control inhibitor. To extract the information for precisely the first pose, the PDBQT files obtained after docking was split using a Python script. For host proteins, the binding affinities of the first poses of all the drugs were compared with the binding affinity of the first pose of the control molecule. The drugs that had an equal or more negative binding energy than that of the control were filtered out and taken forward to be visualized using PyMOL. For viral proteins, it was noticed that the control inhibitors chosen had highly negative binding affinities. Therefore, a cutoff of –7.5 was chosen to select drugs with an equal or more negative binding affinity.

9.4.4.2 PyMOL

PyMOL is a powerful open-source software for visualization scripted in Python, C, and C++ and created by Warren Lyford DeLano. The application, maintained and distributed by Schrödinger, is available for installation on various operating systems like macOS, Windows, Unix, and Linux.

After choosing the most appropriate pose of a docked ligand file, both protein and the ligand (in PDBQT format) were loaded in PyMOL and visualized for the binding pockets of each drug against every target protein. Ligands (from the library), which were able to bind the target protein at the same binding pocket as their control inhibitors, were selected.

9.5 Drug repurposing in the light of COVID-19: Results

9.5.1 *In silico* molecular docking results

The drug repurposing using computational docking methods of 13 differentially expressed host proteins from swab samples in COVID-19-positive versus COVID-19-negative patients was performed and we found five FDA-approved drugs that topped the list by expressing strong binding affinity in more than six of the host proteins. Selinexor (CID 71481097), a drug that has been approved for multiple myeloma and Daunorubicin (CID 30323), an anthracycline antibiotic with antineoplastic activity, inhibited seven proteins each. Figure 9.2b shows the molecular docking of Selinexor with one such target protein, glutathione S-transferase P. Another six proteins were inhibited by Ponatinib (CID 24826799), which has been previously approved for the treatment of chronic myeloid leukemia, and Silmitasertib (CID 24748573) in line for treatment of cholangiocarcinoma, respectively. Dabrafenib (CID 44462760), an antineoplastic agent used in the treatment of metastatic melanoma, showed remarkable results by inhibiting 8 of these 13 biomolecules. Further, it was observed that clinically approved drugs like ABBV-744 (CID 132010322), CPI-0610 (CID 57389999), and Pevonedistat (CID 16720766) inhibited 8 out of 13 proteins.

After performing a similar docking procedure for ten viral proteins (as described in Section 3.4.1), Selinexor and Dabrafenib also displayed inhibition of six of the viral proteins while Ponatinib and Silmitasertib showed these results for four viral proteins. More FDA-approved drugs like Loratadine (CID 3957: an antihistamine), Rapamycin (CID 118701644: an immunosuppressant), and Tigecycline (CID 54686904: a glycylcycline antibiotic) inhibited four viral proteins. Haloperidol (CID 3559: an antipsychotic medication) inhibited five such proteins from the SARS-COV-2 viral proteome. Additionally, clinically approved drugs also made a mark in this aspect, with Pevonedistat and ABBV-744 obtained to inhibit seven and four viral proteins, respectively.

9.5.2 PLIP

PLIP, or the Protein-Ligand Interaction Profiler, is a web application that allows users to gauge the possible interactions between ligand molecules and the proteins they are docked to. It focuses on identifying non-covalent interactions such as hydrogen bonds, π-stacking, and halogen bonds. The server accepts four-character .pdb files without the need for any further preparation. The PLIP server uses OpenBabel to read and process the structures. As mentioned earlier, the shortlisted proteins were then fed to this Profiler. Table 9.4 shows the interacting residues identified using PLIP for glutathione S-transferase P.

9.6 Authors' perspective

9.6.1 Why repurpose existing drugs?

Drug repurposing term was first discussed by Ashburn and Thor back in 2004, who precisely pointed out the enormous burden of a decrease in productivity despite increasing R&D spending (Ashburn and Thor 2004). Such hurdles made modern drug developers come up with solutions with existing drugs, especially

TABLE 9.4

List of Interacting Residues (PLIP) of GST-1 with Top Five Small Molecules

Drugs (FDA Approved)	Initial Function	Hydrogen Bonds	Hydrophobic Interactions	Residues Interacting
Selinexor 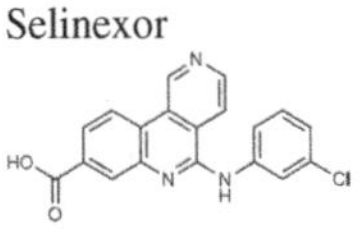	CK2 inhibitor, $IC_{50} = 1$	5	5	8PHE, 7TYR, 38TRP, 51GLN, 52LEU, 108TYR
Dabrafenib 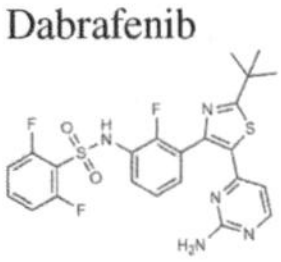	RIPK1 inhibitor, $IC_{50} = 12$	3	3	8PHE, 35VAL, 38TRP, 39GLN, 205GLY
Silmitasertib 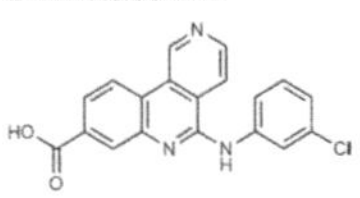	XPO1/nuclear export inhibitor, $IC_{50} = 36$	1	5	8PHE, 104ILE 108TYR, 7TYR
Daunorubicin 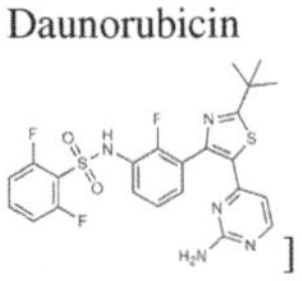	Topoisomerase inhibitor, $Ki = 70$	0	6	8PHE, 38TRP, 104ILE, 108TYR
Ponatinib 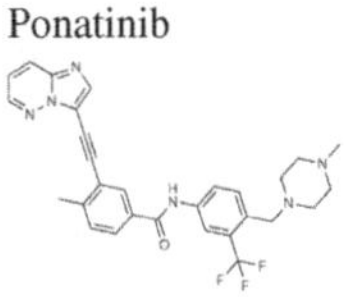	RIPK1 inhibitor, $IC_{50} = 12$	1	5	8PHE, 104ILE, 108TYR, 7TYR

to address the huge timeline and cost issue of conventional drug discovery programs (Vanhaelen et al. 2017). This approach offers us a better trade-off in terms of risk versus reward compared to other drug development strategies. Since then, the field is growing rapidly, and today, a major part of it uses computational approaches (*in silico* experiments) to determine the lead candidates to be taken for further clinical trials.

9.6.2 Drug repurposing as compared to conventional drug discovery and development

The conventional drug discovery process is a time-consuming venture. It takes 10–15 years for a drug since its discovery to get approval for human use. The conventional drug discovery approach results in the identification of new molecular therapeutics. The process consists of five stages: (1) discovery and preclinical phase, (2) safety review phase, (3) clinical research phase, (4) FDA review phase, and (5) post-market safety monitoring. All the stages make the overall process time consuming, costly, with an increased risk of failure. While the drug repurposing approach has only four stages: (1) compound identification, (2) compound acquisition, (3) development stage, and (4) post-market safety monitoring. Advancements in cheminformatics tools supported by biological and structural databases have significantly reduced the time and cost of the drug development process. *In silico* molecular docking techniques with structure-based drug design (SBDD) have further improved the drug repurposing technique. The drug repurposing process is beneficial for developing drugs for rapidly emerging infectious diseases like COVID-19. With conventional drug development, it will take 10–16 years to develop a new drug. On the other hand, this time can be reduced to three to ten years with a drug repurposing approach. With the advancement in omics technologies (genomics, proteomics, interactomics, etc.) and bioinformatics tools, it is surprisingly easy to identify the therapeutic targets for drug repurposing which further simplifies the complex drug discovery process.

9.6.3 Polypharmacology: A new era of computer-aided drug repurposing

Computer-aided drug repurposing uses different evidence to build a study hypothesis that could be clinical and molecular data derived and literature derived (Brown and Patel 2017). Recently, we have been introduced to "Polypharmacology", which focuses on multi-target drugs (MTDs), and is an extended field of *in silico* drug repurposing approach (Lavecchia and Cerchia 2016). It is based on the idea that one drug molecule can interact with many proteins. The results can be both positive (amplifying the drug inhibition) and negative (adverse effect on off-target proteins). However, "Polypharmacology" has proved to be much more beneficial and in the last few years, FDA (Food and Drug Administration, USA) has approved several multi-target drugs for various treatments (Peters 2013). We believe that applying the same approach can be very useful in finding therapeutic alternatives of SARS CoV-2. Proteins from several highly modulated pathways (such as inflammation, neutrophil degranulation, and complement cascade) can be targeted with potential MTDs.

9.6.4 Importance of drug repurposing in the face of infectious disease outbreaks such as the current COVID-19 pandemic

Currently, we are in the middle of the COVID-19 pandemic caused by the SARS-CoV-2 virus and there is no effective drug that can target the virus. This situation suggests the urgent need for a repurposed drug for SARS-Co-V-2 as the *de novo* drug discovery process will take 10–15 years to develop a new drug. Since the beginning of the COVID-19 outbreak, the drug repurposing process has gained considerable momentum to identify a repurposed drug for the life-threatening condition of COVID-19 (Mohamed and Rezaei 2020). Remdesivir is one such example of a repurposed drug, it is a nucleotide analog that inhibits viral RNA polymerase (Grein et al. 2020). It has shown *in vitro* antiviral efficacy against SARS-CoV-2 and was approved in October 2020 to treat COVID-19 (Bellera et al. 2020).

Drug repurposing approaches have also been explored to find drugs for other viral infections as well. Zika virus, an RNA virus of the arbovirus family, caused a large outbreak in Latin America and WHO declared the ZIKV as a public health emergency in 2016 (Wikan and Smith 2016). Mycophenolic acid (Barrows et al. 2016) and Daptomycin (G. Liu et al. 2017) drugs are the repurposed drugs that were approved for the treatment of ZIKV. Ebola virus which belongs to the family of filovirus has caused several outbreaks in past, in 2014–2016 due to its widespread in non-endemic geographical areas, it became a matter of international health emergency (G. Liu et al. 2017). Several repurposed drugs were evaluated for their antiviral efficacy in the infected patients. Favipiravir which is approved for the treatment of the influenza A virus (Sissoko et al. 2016) and GS-5734 which is an adenosine analog (Sheahan et al. 2017) have been shown to decrease the fatality rate compared to the control (Sweiti et al. 2017). Some studies have shown Grazoprevir, a FDA-approved antiviral drug, for the hepatitis C virus-mediated multiple pathway control via inhibition of viral entry targeting host cell proteins ACE-2, TMPRSS2, and RdRp (Behera et al. 2021). Various other antimalarial drugs have also been studied and experimented against SARS-COV 2 infection and a few of them like the quinolines and artemisinin-based drugs have proven to be effective as they are a class of anti-inflammatory drugs (Firestone et al. 2021). Researchers and clinicians have also worked toward inhibition of specific viral and human cellular pathways using drug repurposing like a study identified ten potential candidates for inhibition of SARS-COV-2 3a channel linked with pro-apoptotic function. This list included various FDA-approved drugs, namely, Capreomycin, Pentamidine, Spectinomycin, Kasugamycin, Plerixafor, Flumatinib, Litronesib, Darapladib, Floxuridine, and Fludarabine (Tomar, Krugliak, and Arkin 2021). These drugs are still in clinical trials to confirm their efficacy against the virus and they are yet to be tested in humans for safety.

9.7 Conclusions

Exploring the antiviral efficacy of the already approved drugs offer plausible alternatives for the development of effective drugs. Repurposed drugs could skip the preclinical and phase 1 trial and can directly go to the phase 2 trials as they are already approved for their safety and toxicity. Drug repurposing has shown massive success in developing many new anticancer drugs and there are a few successful examples of drug repurposing against viral infections and a lot can still be explored in this area. This has been possible due to the advancements in the field of omics and big data as with increasing success in these fields, more and more publicly available databases are getting accumulated, which is extremely useful for such computational studies. Today, researchers around the world are contributing to make a COVID-19-free world and at the same time are making us prepared to combat such threats. This chapter thus summarizes the very first footprint of *in silico* drug repurposing, where given *in silico* predictions need to be validated in a further cell line–based study, followed by animal model validation and toxicity effect determination.

REFERENCES

Aamir, Mohd, Vinay Kumar Singh, Manish Kumar Dubey, Mukesh Meena, Sarvesh Pratap Kashyap, Sudheer Kumar Katari, Ram Sanmukh Upadhyay, Amineni Umamaheswari, and Surendra Singh. 2018. "In Silico Prediction, Characterization, Molecular Docking, and Dynamic Studies on Fungal SDRs as Novel Targets for Searching Potential Fungicides Against Fusarium Wilt in Tomato." *Frontiers in Pharmacology* 9: 1038.

Ahmed, Sabeena, Mohammad Mahbubul Karim, Allen G. Ross, Mohammad Sharif Hossain, John D. Clemens, Mariya Kibtiya Sumiya, Ching Swe Phru, et al. 2021. "A Five-Day Course of Ivermectin for the Treatment of COVID-19 May Reduce the Duration of Illness." *International Journal of Infectious Diseases* 103: 214–16. https://doi.org/10.1016/j.ijid.2020.11.191.

Ashburn, Ted T., and Karl B. Thor. 2004. "Drug Repositioning: Identifying and Developing New Uses for Existing Drugs." *Nature Reviews. Drug Discovery* 3 (8). England: 673–83. https://doi.org/10.1038/nrd1468.

Bankar, Renuka, Kruthi Suvarna, Saicharan Ghantasala, Arghya Banerjee, Deeptarup Biswas, Manisha Choudhury, Viswanthram Palanivel, et al. 2021. "Proteomic Investigation Reveals Dominant Alterations of Neutrophil Degranulation and MRNA Translation Pathways in Patients with COVID-19." IScience 24 (3): 102135. https://doi.org/10.1016/j.isci.2021.102135.

Barh, Debmalya, Sandeep Tiwari, Marianna E. Weener, Vasco Azevedo, Aristóteles Góes-Neto, M. Michael Gromiha, and Preetam Ghosh. 2020. "Multi-Omics-Based Identification of SARS-CoV-2 Infection Biology and Candidate Drugs against COVID-19." *Computers in Biology and Medicine* 126: 104051. https://doi.org/10.1016/j.compbiomed.2020.104051.

Barrows, Nicholas J., Rafael K. Campos, Steven T. Powell, K. Reddisiva Prasanth, Geraldine Schott-Lerner, Ruben Soto-Acosta, Gaddiel Galarza-Muñoz, et al. 2016. "A Screen of FDA-Approved Drugs for Inhibitors of Zika Virus Infection." *Cell Host & Microbe* 20 (2): 259–70. https://doi.org/10.1016/j.chom.2016.07.004.

Behera, Santosh Kumar, Nazmina Vhora, Darshan Contractor, Amit Shard, Dinesh Kumar, Kiran Kalia, and Alok Jain. 2021. "Computational Drug Repurposing Study Elucidating Simultaneous Inhibition of Entry and Replication of Novel Corona Virus by Grazoprevir." *Scientific Reports* 11 (1): 7307. https://doi.org/10.1038/s41598-021-86712-2.

Bellera, Carolina L., Manuel Llanos, Melisa E. Gantner, Santiago Rodriguez, Luciana Gavernet, Marcelo Comini, and Alan Talevi. 2020. "Can Drug Repurposing Strategies Be the Solution to the COVID-19 Crisis?" *Expert Opinion on Drug Discovery*, December. England: 1–8. https://doi.org/10.1080/17460441.2021.1863943.

Bojkova, Denisa, Kevin Klann, Benjamin Koch, Marek Widera, David Krause, Sandra Ciesek, Jindrich Cinatl, and Christian Münch. 2020. "Proteomics of SARS-CoV-2-Infected Host Cells Reveals Therapy Targets." *Nature* 583 (7816): 469–72. https://doi.org/10.1038/s41586-020-2332-7.

Bouhaddou, Mehdi, Danish Memon, Bjoern Meyer, Kris M. White, Veronica V. Rezelj, Miguel Correa Marrero, Benjamin J. Polacco, et al. 2020. "The Global Phosphorylation Landscape of SARS-CoV-2 Infection." *Cell* 182 (3): 685–712. e19. https://doi.org/10.1016/j.cell.2020.06.034.

Brown, Adam S., and Chirag J. Patel. 2017. "A Standard Database for Drug Repositioning." *Scientific Data* 4 (March): 170029. https://doi.org/10.1038/sdata.2017.29.

Caly, Leon, Julian D. Druce, Mike G. Catton, David A. Jans, and Kylie M. Wagstaff. 2020. "The FDA-Approved Drug Ivermectin Inhibits the Replication of SARS-CoV-2 in Vitro." *Antiviral Research* 178: 104787. https://doi.org/10.1016/j.antiviral.2020.104787.

Chen, Yan-Mei, Yuanting Zheng, Ying Yu, Yunzhi Wang, Qingxia Huang, Feng Qian, Lei Sun, et al. 2020. "Blood Molecular Markers Associated with COVID-19 Immunopathology and Multi-Organ Damage." *The EMBO Journal* 39 (24). John Wiley & Sons, Ltd: e105896. https://doi.org/10.15252/embj.2020105896.

D'Alessandro, Angelo, Tiffany Thomas, Monika Dzieciatkowska, Ryan C. Hill, Richard O. Francis, Krystalyn E. Hudson, James C. Zimring, Eldad A. Hod, Steven L. Spitalnik, and Kirk C. Hansen. 2020. "Serum Proteomics in COVID-19 Patients: Altered Coagulation and Complement Status as a Function of IL-6 Level." *Journal of Proteome Research* 19 (11). American Chemical Society: 4417–27. https://doi.org/10.1021/acs.jproteome.0c00365.

Filbin, Michael R., Arnav Mehta, Alexis M. Schneider, Kyle R. Kays, Jamey R. Guess, Matteo Gentili, Bánk G. Fenyves, et al. 2020. "Plasma Proteomics Reveals Tissue-Specific Cell Death and Mediators of Cell-Cell Interactions in Severe COVID-19 Patients." *BioRxiv*, January. https://doi.org/10.1101/2020.11.02.365536.

Firestone, Tessa M., Opeoluwa O. Oyewole, St Patrick Reid, and Caroline L. Ng. 2021. "Repurposing Quinoline and Artemisinin Antimalarials as Therapeutics for SARS-CoV-2: Rationale and Implications." *ACS Pharmacology & Translational Science*, March. American Chemical Society. https://doi.org/10.1021/acsptsci.0c00222.

Frantzi, Maria, Agnieszka Latosinska, and Harald Mischak. 2019. "Proteomics in Drug Development: The Dawn of a New Era?" *Proteomics. Clinical Applications* 13 (February): 1800087. https://doi.org/10.1002/prca.201800087.

Gao, Yan, Liming Yan, Yucen Huang, Fengjiang Liu, Yao Zhao, Lin Cao, Tao Wang, et al. 2020. "Structure of the RNA-Dependent RNA Polymerase from COVID-19 Virus." *Science* 368 (6492): 779LP–82. https://doi.org/10.1126/science.abb7498.

Gordon, David E., Gwendolyn M. Jang, Mehdi Bouhaddou, Jiewei Xu, Kirsten Obernier, Kris M. White, Matthew J. O'Meara, et al. 2020. "A SARS-CoV-2 Protein Interaction Map Reveals Targets for Drug Repurposing." *Nature* 583 (7816): 459–68. https://doi.org/10.1038/s41586-020-2286-9.

Grein, Jonathan, Norio Ohmagari, Daniel Shin, George Diaz, Erika Asperges, Antonella Castagna, Torsten Feldt, et al. 2020. "Compassionate Use of Remdesivir for Patients with Severe Covid-19." *The New England Journal of Medicine* 382 (24): 2327–36. https://doi.org/10.1056/NEJMoa2007016.

Hodos, Rachel A., Brian A. Kidd, Khader Shameer, Ben P. Readhead, and Joel T. Dudley. 2016. "In Silico Methods for Drug Repurposing and Pharmacology." *WIREs Systems Biology and Medicine* 8 (3). John Wiley & Sons, Ltd: 186–210. https://doi.org/10.1002/wsbm.1337.

Hopkins, Andrew L., and Colin R. Groom. 2002. "The Druggable Genome." *Nature Reviews Drug Discovery* 1 (9): 727–30. https://doi.org/10.1038/nrd892.

Hou, Xin, Xiaomei Zhang, Xian Wu, Minya Lu, Dan Wang, Meng Xu, Hongye Wang, et al. 2020. "Serum Protein Profiling Reveals a Landscape of Inflammation and Immune Signaling in Early-Stage COVID-19 Infection." *Molecular & Cellular Proteomics* 19 (11): 1749–59. https://doi.org/10.1074/mcp.RP120.002128.

Hung, Ivan Fan-Ngai, Kwok-Cheung Lung, Eugene Yuk-Keung Tso, Raymond Liu, Tom Wai-Hin Chung, Man-Yee Chu, Yuk-Yung Ng, et al. 2020. "Triple Combination of Interferon Beta-1b, Lopinavir—Ritonavir, and Ribavirin in the Treatment of Patients Admitted to Hospital with COVID-19: An Open-Label, Randomised, Phase 2 Trial." *The Lancet* 395 (10238): 1695–704. https://doi.org/10.1016/S0140-6736(20)31042-4.

Kong, Ren, Feng Wang, Jian Zhang, Fengfei Wang, and Shan Chang. 2019. "CoDockPP: A Multistage Approach for Global and Site-Specific Protein—Protein Docking." *Journal of Chemical Information and Modeling* 59 (8). American Chemical Society: 3556–64. https://doi.org/10.1021/acs.jcim.9b00445.

Kuzmanov, Uros, and Andrew Emili. 2013. "Protein-Protein Interaction Networks: Probing Disease Mechanisms Using Model Systems." *Genome Medicine* 5 (4): 37. https://doi.org/10.1186/gm441.

Lavecchia, Antonio, and Carmen Cerchia. 2016. "In Silico Methods to Address Polypharmacology: Current Status, Applications and Future Perspectives." *Drug Discovery Today* 21 (2). England: 288–98. https://doi.org/10.1016/j.drudis.2015.12.007.

Leng, Ling, Ruiyuan Cao, Jie Ma, Danlei Mou, Yunping Zhu, Wei Li, Luye Lv, et al. 2020. "Pathological Features of COVID-19-Associated Lung Injury: A Preliminary Proteomics Report Based on Clinical Samples." *Signal Transduction and Targeted Therapy* 5 (1): 240. https://doi.org/10.1038/s41392-020-00355-9.

Li, Na, Lingfeng Zhao, and Xianquan Zhan. 2021. "Quantitative Proteomics Reveals a Broad-Spectrum Antiviral Property of Ivermectin, Benefiting for COVID-19 Treatment." *Journal of Cellular Physiology* 236 (4). John Wiley & Sons, Ltd: 2959–75. https://doi.org/10.1002/jcp.30055.

Liu, Guodong, Gary Wong, Shuo Su, Yuhai Bi, Frank Plummer, George F. Gao, Gary Kobinger, and Xiangguo Qiu. 2017. "Clinical Evaluation of Ebola Virus Disease Therapeutics." *Trends in Molecular Medicine* 23 (9): 820–30. https://doi.org/10.1016/j.molmed.2017.07.002.

Liu, Xiaoyan, Zhe Li, Shuai Liu, Jing Sun, Zhanghua Chen, Min Jiang, Qingling Zhang, et al. 2020. "Potential Therapeutic Effects of Dipyridamole in the Severely Ill Patients with COVID-19." *Acta Pharmaceutica Sinica B* 10 (7): 1205–15. https://doi.org/10.1016/j.apsb.2020.04.008.

Luo, Pan, Yi Liu, Lin Qiu, Xiulan Liu, Dong Liu, and Juan Li. 2020. "Tocilizumab Treatment in COVID-19: A Single Center Experience." *Journal of Medical Virology* 92 (7). John Wiley & Sons, Ltd: 814–18. https://doi.org/10.1002/jmv.25801.

Martin, Roman, Hannah F. Löchel, Marius Welzel, Georges Hattab, Anne-Christin Hauschild, and Dominik Heider. 2020. "CORDITE: The Curated CORona Drug InTERactions Database for SARS-CoV-2." *IScience* 23 (7): 101297. https://doi.org/10.1016/j.isci.2020.101297.

Messner, Christoph B., Vadim Demichev, Daniel Wendisch, Laura Michalick, Matthew White, Anja Freiwald, Kathrin Textoris-Taube, et al. 2020. "Ultra-High-Throughput Clinical Proteomics Reveals Classifiers of COVID-19 Infection." *Cell Systems* 11 (1): 11–24.e4. https://doi.org/10.1016/j.cels.2020.05.012.

Mohamed, Kawthar, and Nima Rezaei. 2020. "COVID-19 Pandemic Is Not the Time of Trial and Error." *The American Journal of Emergency Medicine* 46: 774–75. https://doi.org/10.1016/j.ajem.2020.09.020.

Peters, Jens-Uwe. 2013. "Polypharmacology—Foe or Friend?" *Journal of Medicinal Chemistry* 56 (22). United States: 8955–71. https://doi.org/10.1021/jm400856t.

Schmidt, Nora, Caleb A. Lareau, Hasmik Keshishian, Sabina Ganskih, Cornelius Schneider, Thomas Hennig, Randy Melanson, et al. 2021. "The SARS-CoV-2 RNA—Protein Interactome in Infected Human Cells." *Nature Microbiology* 6 (3): 339–53. https://doi.org/10.1038/s41564-020-00846-z.

Sheahan, Timothy P., Amy C. Sims, Rachel L. Graham, Vineet D. Menachery, Lisa E. Gralinski, James B. Case, Sarah R. Leist, et al. 2017. "Broad-Spectrum Antiviral GS-5734 Inhibits Both Epidemic and Zoonotic Coronaviruses." *Science Translational Medicine* 9 (396). https://doi.org/10.1126/scitranslmed.aal3653.

Shen, Bo, Xiao Yi, Yaoting Sun, Xiaojie Bi, Juping Du, Chao Zhang, Sheng Quan, et al. 2020. "Proteomic and Metabolomic Characterization of COVID-19 Patient Sera." *Cell* 182 (1): 59–72.e15. https://doi.org/10.1016/j.cell.2020.05.032.

Sissoko, Daouda, Cedric Laouenan, Elin Folkesson, Abdoul-Bing M'Lebing, Abdoul-Habib Beavogui, Sylvain Baize, Alseny-Modet Camara, et al. 2016. "Experimental Treatment with Favipiravir for Ebola Virus Disease (the JIKI Trial): A Historically Controlled, Single-Arm Proof-of-Concept Trial in Guinea." *PLoS Medicine* 13 (3): e1001967. https://doi.org/10.1371/journal.pmed.1001967.

Stone, John H., Matthew J. Frigault, Naomi J. Serling-Boyd, Ana D. Fernandes, Liam Harvey, Andrea S. Foulkes, Nora K. Horick, et al. 2020. "Efficacy of Tocilizumab in Patients Hospitalized with Covid-19." *New England Journal of Medicine* 383 (24). Massachusetts Medical Society: 2333–44. https://doi.org/10.1056/NEJMoa2028836.

Su, Yapeng, Daniel Chen, Dan Yuan, Christopher Lausted, Jongchan Choi, Chengzhen L. Dai, Valentin Voillet, et al. 2020. "Multi-Omics Resolves a Sharp Disease-State Shift between Mild and Moderate COVID-19." *Cell* 183 (6): 1479–95.e20. https://doi.org/10.1016/j.cell.2020.10.037.

Suvarna, Kruthi, Deeptarup Biswas, Medha Gayathri J. Pai, Arup Acharjee, Renuka Bankar, Viswanthram Palanivel, Akanksha Salkar, et al. 2021. "Proteomics and Machine Learning Approaches Reveal a Set of Prognostic Markers for COVID-19 Severity With Drug Repurposing Potential." Frontiers in Physiology 12: 652799. https://doi.org/10.3389/fphys.2021.652799.

Sweiti, Hussein, Obinna Ekwunife, Thomas Jaschinski, and Stefan K. Lhachimi. 2017. "Repurposed Therapeutic Agents Targeting the Ebola Virus: A Systematic Review." *Current Therapeutic Research, Clinical and Experimental* 84: 10–21. https://doi.org/10.1016/j.curtheres.2017.01.007.

Tomar, Prabhat Pratap Singh, Miriam Krugliak, and Isaiah T. Arkin. 2021. "Blockers of the SARS-CoV-2 3a Channel Identified by Targeted Drug Repurposing." *Viruses* 13 (3). Switzerland. https://doi.org/10.3390/v13030532.

Valle, Diane Marie Del, Seunghee Kim-Schulze, Hsin-Hui Huang, Noam D. Beckmann, Sharon Nirenberg, Bo Wang, Yonit Lavin, et al. 2020. "An Inflammatory Cytokine Signature Predicts COVID-19 Severity and Survival." *Nature Medicine* 26 (10): 1636–43. https://doi.org/10.1038/s41591-020-1051-9.

Vanhaelen, Quentin, Polina Mamoshina, Alexander M. Aliper, Artem Artemov, Ksenia Lezhnina, Ivan Ozerov, Ivan Labat, and Alex Zhavoronkov. 2017. "Design of Efficient Computational Workflows for in Silico Drug Repurposing." *Drug Discovery Today* 22 (2). England: 210–22. https://doi.org/10.1016/j.drudis.2016.09.019.

Verma, Amit, Amitava Adhikary, Gayle Woloschak, Bilikere S. Dwarakanath, and Rao V. L. Papineni. 2020. "A Combinatorial Approach of a Polypharmacological Adjuvant 2-Deoxy-D-Glucose with Low Dose Radiation Therapy to Quell the Cytokine Storm in COVID-19 Management." *International Journal of Radiation Biology* 96 (11). Taylor & Francis: 1323–28. https://doi.org/10.1080/09553002.2020.1818865.

Wikan, Nitwara, and Duncan R. Smith. 2016. "Zika Virus: History of a Newly Emerging Arbovirus." *The Lancet. Infectious Diseases* 16 (7). United States: e119–26. https://doi.org/10.1016/S1473-3099(16)30010-X.

Xu, Jun, and Arnold Hagler. 2002. "Chemoinformatics and Drug Discovery." *Molecules* 7 (8): 566–600. https://doi.org/10.3390/70800566.

10

Spectroscopy methods for SARS-CoV-2 detection

Abhiram Gokhale, Ashwini Khaladkar, and Arghya Banerjee

CONTENTS

List of abbreviations

ATR-FTIR – Attenuated total reflection- Fourier transform infrared spectroscopy
AUROC – Area under the receiver operating characteristic curve
CCD – Charge-coupled device
COVID-19 – Coronavirus Disease 2019
CRM – Confocal Raman Microscopy
CVF – Cerebrospinal fluid
DENV – Dengue virus
GA-LDA – Genetic algorithm-linear discriminant analysis
GFP – Green fluorescent protein
HBV – Hepatitis B virus
HCV – Hepatitis C Virus
HIV – Human immunodeficiency virus
HPV – Human papilloma virus
HSV – Herpes simplex virus
MDCK – Madin-Darby canine kidney
PLSDA – Partial least squares discriminant analysis
PPV – Porcine parvovirus
RSV – Respiratory syncytial virus
RT-PCR – Real time Polymerase chain reaction
SARS-CoV-2 – Severe acute respiratory syndrome coronavirus 2

DOI: 10.1201/9781003220787-10

SD-IR – Second derivative infrared
SVM – Support vector machine
TYMV – Turnip yellow mosaic virus
VPH – volume-phase holographic
VTM – Viral transport medium
VZV – Varicella zoster viruses

10.1 Introduction

Various spectroscopy approaches can provide rapid, accurate, and relatively cost-effective methods to address various limitations of gold standard RT-PCR tests. The basic principle of spectroscopy is based on the interplay of electromagnetic radiation with a medium or the measurement of radiation emitted from the medium. An average spectrometer includes a light source, so when the samples are exposed to light, the sample molecules selectively absorb radiation of specific wavelengths. This absorption instigates the sample molecules to transition from the ground state to the excited state. The detectors identify these differences between the states of vibrational energy and installed software in the computer calculates it as the absorption peak frequency and give readout. Historically, these techniques have been routinely used to detect and characterize chemical compounds in bioengineering, natural sciences, and now in the medical field. Vibrational spectroscopy, including attenuated total reflection Fourier transform infrared (ATR-FTIR) spectroscopy, and Raman spectroscopy, has been routinely used to discriminate and classify normal and pathological populations using biospecimens (see Tables 10.1 and 10.2).

Both techniques have been extensively used in clinics for diagnostic purposes. Based on the spectra recorded from the patient's complex biological samples such as urine, CSF, blood, saliva, breast milk, and plasma serum, it helps for diagnostics of neurodegenerative diseases (Nabers et al. 2018), diabetes (Pleitez et al. 2013), and cancer (Auner et al. 2018; Lasch et al. 2004; Kallenbach-Thieltges et al. 2013). These techniques have regularly been used for the detection of viral signature in infected patients in cases of HIV, Zika, herpes, influenza, etc.(Salman et al. 2002; Manoto et al. 2021; Yeh et al. 2020). This chapter will discuss the working principle of FTIR and Raman spectroscopy for the SARS-CoV-2 detection. Herein we will discuss bio specimen collection from the patient and how the collected samples were processed for the detection of SARS-CoV-2. Subsequently, we will discuss how the spectral readings in FTIR and Raman spectroscopy are obtained and analyzed for a fast, efficient way to detect SARS-CoV-2.

10.2 Working principle of FTIR and Raman

Originally infrared spectrometers were "dispersive", which means prisms or graters were used to separate the individual frequencies of energy emitted from the infrared source. After passing the radiation through a sample, the amount of energy at each frequency is measured by a detector. This method took a considerable amount of

TABLE 10.1

Recent Milestones of Work Done in Viral Detection Using Raman

Study	Sample Used	Aim	Year
SARS-CoV-2	Human sera	Efficient primary screening of COVID-19 by serum Raman spectroscopy	2021
RNA viruses	Human saliva	Raman spectroscopy-based detection of RNA viruses in saliva: a preliminary report	2020
TYMV infection	Virus-infected leaves extract	Turnip yellow mosaic virus infection in Chinese cabbage plant	2013
Herpes virus types (HSV)-1, HSV-2, and VZV infection	Virus-infected vero cells	Vero cells infected with herpes simplex virus (HSV) types (HSV1 and HSV2) and varicella zoster viruses	2014
HIV infection	Plasma	HIV-1 virus	2017
Liver disease progression from steatosis to fibrosis	Biopsy	Liver lesions induced by hepatitis C virus	2015
DENV infection	Serum	Lactate based on optical detection of dengue virus infection in human serum	2017
Hepatitis B virus infection	Serum	Analysis of serum for hepatitis B virus (HBV) infection	2018
Typhoid and dengue infection	Serum	Differentiation of typhoid and dengue fever based on sera obtained from infected humans	2018
HBV infection	Serum	Hepatitis B virus (HBV) infection	2019
Virus-elicited metabolic reactions	Virus-infected MDCK cells	Metabolic study of influenza A virus induced in mammalian cells	2020

TABLE 10.2

Recent Milestones of Work Done in Viral Detection Using FTIR

Study	Sample Used	Aim	Year
SARS-CoV-2	Human swab	Ultra rapid on-site detection of SARS-CoV-2 infection using simple ATR-FTIR spectroscopy and an analysis algorithm: high sensitivity and specificity	2021
SARS-CoV-2	Human sera	Fast screening and primary diagnosis of COVID-19 by ATR-FTIR	2021
Polio virus	Buffalo green monkey kidney cells	Detection and quantification of poliovirus infection	2011

(Continued)

TABLE 10.2 (Continued)

Study	Sample Used	Aim	Year
Hepatitis B and C virus	Human sera	Diagnosis of hepatitis B and C virus	2019
HPV	Cervical scraps	HPV detection for cervical cancer	2018
HIV and HCV	Plasma	Differentiating HIV-infected patients with HIV HCV co- infection	2017
Herpes virus HSV VZV	Vero, human fibroblast, rabbit fibroblast cell lines	Detection of virus-infected cells	2002
Dengue virus	Freeze-dried human sera	Dengue diagnosis	2019

time, and FTIR was designed to overcome this disadvantage. FTIR uses an interferometer to overcome this disadvantage by measuring all infrared frequencies simultaneously.

In FTIR spectroscopy, the energy from the infrared source goes to the beam splitter, which splits the beam into two parts. The first part of the beam is transmitted to a moving mirror, and the other is reflected in a fixed mirror. The moving mirror oscillates at a constant velocity. An interference pattern is created at the beam splitter where these beams converge after traveling different distances at different velocities; some of the wavelengths recombine constructively and some destructively. This interference pattern is called an interferogram. This interferogram from the beam splitter passes through the sample (which is made of biomolecules with an electric dipole), where some energy is absorbed, and some are transmitted. The transmitted portion reaches the detector. The detector reads information about every wavelength in the infrared range simultaneously. Subsequently, the data from the detector is Fourier transformed using computational tools, and spectra are presented as a function of wavelength and intensity. Examination of samples in a solid and liquid state is now possible with FTIR in conjugation with attenuated total reflection (ATR). In general, the beam of infrared light is passed through the ATR crystal leading to the formation of an evanescent wave due to total internal reflection and extends to the samples to be analyzed (Francis M Mirabella 1993) (Figure 10.1).

Similarly, in Raman spectroscopy, when light is passed through the sample molecules, the incident light is scattered, absorbed, or transmitted. The majority of scattered light is elastically scattered, and this type of scattering is referred to as *Rayleigh scattering*, where the energy of the scattered light is equal to that of the incident light; but when these energies are not equal, it results in inelastic scattering. Inelastic scattering can be of two types: Stokes scattering and anti-Stokes scattering. These terminologies are defined based on molecular vibrational energies. The difference in energy is equal to the difference in energies of vibrational energy levels within the ground electronic state. When the energy (or frequency) of the scattered photon is less than that of the incident photon, it is known as Stokes scattering, and if it is greater, then it is known as Anti-stokes scattering. Raman spectrum is a measurement of Raman shift which is the difference in energy of incident light and scattered light. The shift found in the angular frequency of the scattered light can be explained via the following equation:

$$\omega_{scat} = \omega_p \pm \omega_{osc}$$

Where subscript osc is the molecular vibration, p is the incident photon, and scat is the scattered light (Duncan, Reintjes, and Manuccia 1982). Raman spectroscopy is used to perceive rotational, vibrational, and other low-frequency modes. It uses the interaction of the photon with matter to understand the signature information of the molecule. It depends on scattered light and gives information about intramolecular and intermolecular vibrations of the sample molecules.

One of the improved versions of Raman spectroscopy in the combination of microscopy CRM (confocal Raman Microscopy) has been advantageous in revealing information about the depth and buried structures that are present in thin samples with a high spatial resolution (Giridhar, Manepalli, and Apparao 2017) (Figure 10.1).

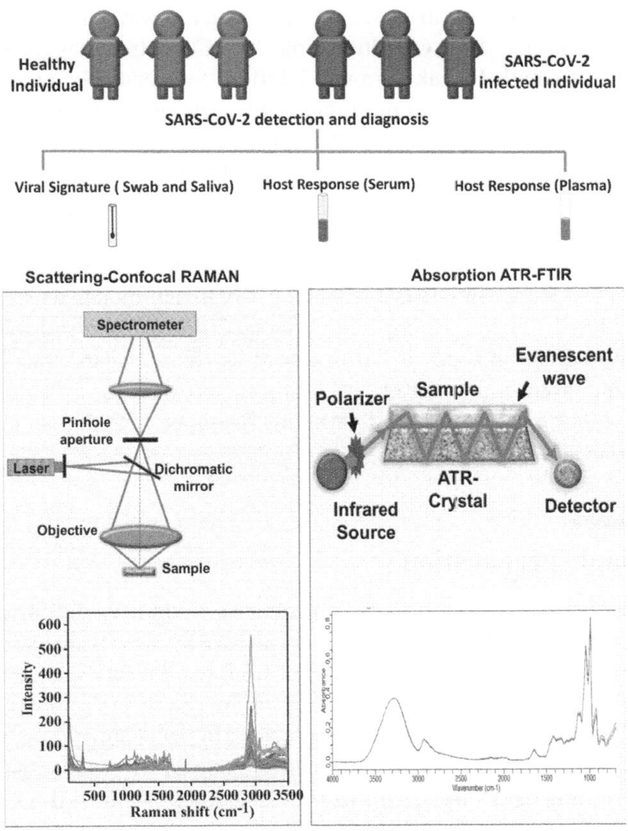

FIGURE 10.1 General schematic workflow of analyzing clinical samples using ATR-FTIR and confocal Raman spectroscopy.

10.3 Fingerprint region

Spectra obtained from FTIR detectors and from Raman spectroscopy are the characteristic features of the sample containing biomolecules such as proteins, lipids, carbohydrates, and DNA. These infrared spectra are represented on a Cartesian plane where the X-axis (peak position) represents the frequency of vibration of a specific part of the molecule (4,000 to 400 cm^{-1}) and the Y-axis (peak intensity) informs about the absorbed sample energy. Each of the biomolecules mentioned previously has a specific signature region on the X-axis called the "bio fingerprint region" which enables the discrimination of the different populations and identification of potential biomarkers. For interrogating biological materials, the most important spectral regions measured are typically the fingerprint region (600–1,450 cm^{-1}) and the Amide I and amide II (Amide I/II) regions (1,500–1,700 cm^{-1}). The higher wavenumber region (2,550–3,500 cm^{-1}) is associated with stretching vibrations such as S–H, C–H, N–H, and O–H, whereas the lower wavenumber regions typically correspond to bending and carbon skeleton vibrations (Khan and Rehman 2020). Proteins mainly contribute to the Amide I peak at wavenumber around 1,650 cm^{-1}, which comprises 80% CO_2 stretching, 10% C–N stretching, and 10% N–H bending and the Amide II peak at around 1,540 cm^{-1} wavenumber due to 60% N–H bending and 40% C–N stretching. Lipids contribute absorptions at ~1,740 cm^{-1} wavenumber, PO_2 stretching vibrations of DNA are located at ~1,080 cm^{-1} (symmetric; $nsPO_2$) and ~1,225 cm^{-1} (asymmetric; $nasPO_2$).

In the case of biological samples, approximately 90% of the peaks seen in the Raman spectrum are in the "fingerprint" spectral region containing Δv ~500–1,800 cm^{-1} (Δv is the wavenumber shift usually expressed in cm^{-1}), with the remaining found at higher energy in CH/OH stretching vibrational modes which are covering Δv ~2,700 –3,300 cm^{-1} (Camp and Cicerone 2015). Raman signature range is attributable to RNA moieties of viruses such as ribose sugars: 900–1,045 cm^{-1}, the phosphate group: 1,080–1,100 cm^{-1}, nitrogenous uracil base: 780 cm^{-1}, ribose phosphate: 1,044 cm^{-1}, A/G ring: 1,480 cm^{-1} (Desai et al. 2020; De Gelder et al. 2007; Gong et al. 2009).

10.4 Sample preparation

Zhang et al. in 2021 analyzed the serum samples using it for fast screening and primary diagnostics of the patients infected with SARS-CoV-2 using the ATR-FTIR system (Zhang et al. 2021). Blood collected from the patients was centrifuged at 5 000 rpm for 5 minutes for obtaining the serum. The serum samples were heat-inactivated to deactivate the potential pathogens by incubating them at 56°C for 30 minutes. PerkinElmer infrared spectrometer coupled with a diamond ATR was used for recording the FTIR spectra with a resolution of 4 cm^{-1}. Around 3–4 μl of the sample was loaded on the ATR crystal and was dried under mild airflow at room temperature. Spectrum for each sample was recorded one to three times with an accumulation of 16 scans per spectrum. After a period of 5 minutes when the sample was dried, spectra were recorded between 4,000 and 600 cm^{-1}, water absorption was monitored for OH peak stretching around 3,300 cm^{-1}, and bending around 1,635 cm^{-1} with the spectral background recorded separately to achieve a high signal-to-noise ratio. The constant baseline drift observed at 4,000 cm^{-1} was subtracted, followed by normalization with Amide I absorbance. For each normalized spectra, the second derivative infrared (SD-IR) was calculated (Zhang et al. 2021).

Barauna et al. in 2021 analyzed the swab samples for the on-site detection of SARS-CoV-2 using ATR-FTIR techniques with high specificity and sensitivity (Barauna et al. 2021). Pharyngeal swab from the patient was collected with the help of a cotton swab from the inner part of the mouth by scrapping the tonsil area, cheek, and tongue after absorbing the secretory fluids. The swab was placed in a sterile tube and was stored at 4°C until spectral reading. The spectra of swab samples were recorded with Agilent Cary 630 FTIR spectrometer. Each clinical sample was recorded in triplicates from a range of 4,000 to 650 cm^{-1} with each spectrum containing 1,798 points with a resolution of 1.86 cm^{-1} by directly placing the swab samples in the crystal of Agilent Cary 630 FTIR spectrometer. Each spectrum was scanned for 32 coadditions, interspersed with 32 background scans. After the scans were complete, the crystal was wiped with MilliQ water followed by 70% ethanol to avoid any cross-contamination across the sample (Barauna et al. 2021).

In most cases, swabs are mainly collected in VTM tubes. To study the efficacy swab samples collected in the VTM tubes that can be used for AT-FTIR studies, nasopharyngeal swabs were collected following the ICMR guidelines. The samples were collected by trained medical officers in a COVID-19 designated hospital. The swabs were stored in tubes containing Viral Transport Media (VTM) (CDC: SOP#: DSR-052–05) at 4°C overnight at the hospital. Different approaches were followed for sample preparation so as to select the optimal protocol for spectral reading. In a tube, 100 μl of VTM was mixed with 300 μl of pre- chilled ethanol and vortexed for 2–3 minutes and kept for overnight incubation at –20°C. On a subsequent day, the tube was centrifuged at 12,000 *g* for 20 minutes at 4°C, and the supernatant was used for the spectral reading. In the second protocol, after the overnight incubation, the upper layer of the bilayer was used for spectral reading. In the third protocol, the samples were vortexed following the overnight incubation; spectral reading was then recorded. FTIR spectra were recorded on the Cary 630 (Agilent) spectrometer coupled with the ATR module. The ATR detection diamond was cleaned and background spectra were acquired before each acquisition with 100% v/v ethanol. On the crystal was placed 1 μl of sample and air-dried. Each sample was recorded as technical duplicates. Based on the results (Figure 10.2), protocol 3, where samples were vortexed following overnight incubation, showed differential spectra and can be used for analysis. One should always keep in mind that if the swab is collected in the VTM tubes, VTM constituents such as protein (serum, albumin, or gelatin) and antibiotics will interfere with the ATR-FTIR spectral annotation.

Desai et al. in 2020 studied the detection of RNA viruses in saliva using Raman spectroscopy (Yin et al. 2021); they detected the presence of RNA virions, which was generated using pLL 3.7 lentivirus vector system (based on HIV-1). These virion particles were spiked in water and saliva from the stock of 7.05 × 107 TU/ml at a dilution of 1:10 and 1:1,000, respectively. GFP expression in HEK293FT

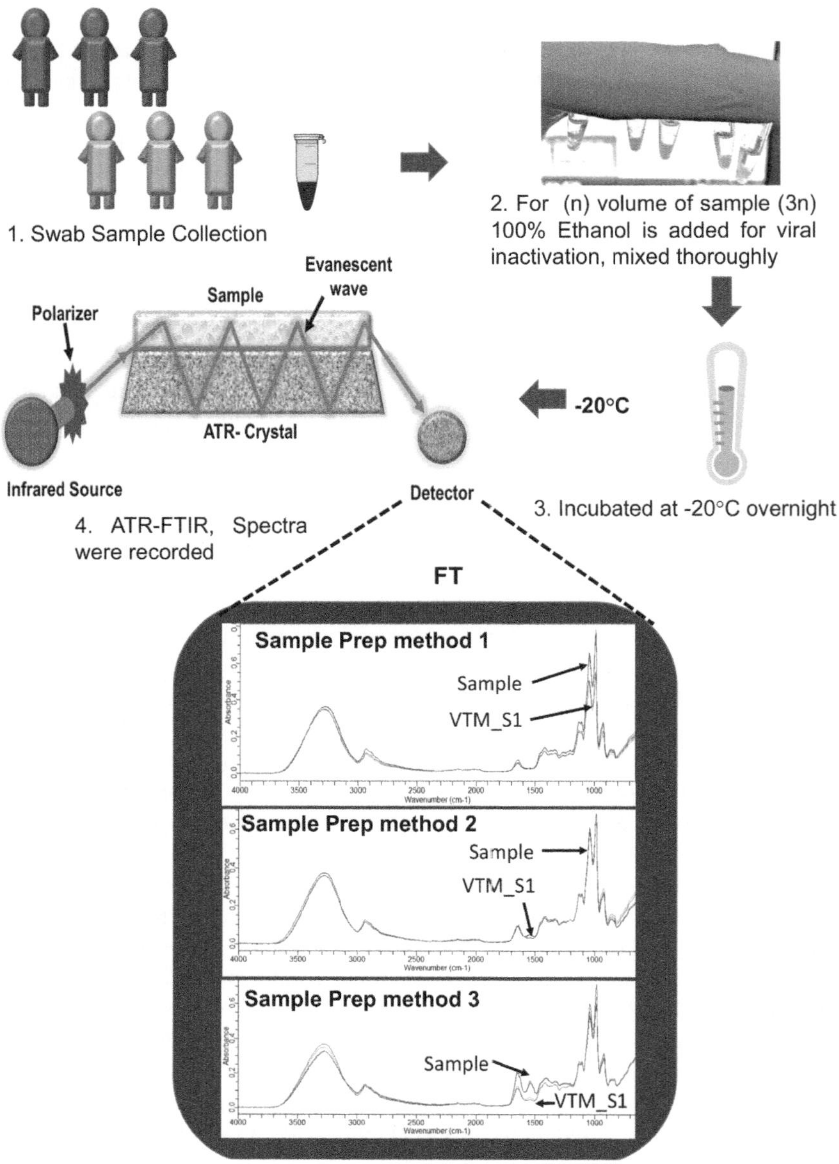

FIGURE 10.2 Schematic workflow analyzing SARS-CoV-2 swab samples using ATR-FTIR.

cells was used for the confirmation of infectivity of viral particles. Thus, mimicking the patient's saliva infected with viral particles, around 20 µl of the sample was smeared on CaF2 slides and allowed to air-dry. Raman spectrum was then recorded on a Raman spectrometer.

In an experimental design, plasma samples of SARS-CoV-2 infected patients were used for the study. Blood from infected patients was collected from the patients by trained medical officers in a COVID-19 designated hospital. Around 5 ml of blood was collected in EDTA vacutainer by clinicians from confirmed patients following the infection control protocol approved by WHO and ICMR. Plasma was isolated by centrifuging blood at 10,000 rpm for 5 minutes. Supernatant plasma was taken for clinical diagnostics purposes, and leftover plasma samples were aliquoted into cryovials (~1 ml) and stored at −80°C for studies. Plasma was then diluted in the ratio of 1:3 with 100% v/v chilled ethanol. Subsequently, thawed and mixed 10 μl of plasma is smeared on CaF2 slides and allowed to air-dry, followed by confocal Raman. The settings used in Raman spectrometer were as follows: excitation wavelength (nm): 532.123; laser power (mW): 5.538; central wavelength (nm): 602.677; number of accumulations: 10; integration time (s): 10; objective name: Zeiss EC Epipian 20x/0.4; objective magnification: 20. Schematic representational workflow can be seen. Ultimately, these obtained Raman spectra have been used for ahead data analysis (Yin et al. 2021) (Figure 10.3).

Yin et al. in 2021 studied SARS-CoV-2 infected serum samples for the primary screening of SARS-CoV-2 (Yin et al. 2021). Blood samples were collected from SARS-CoV-2 infected patients; serum samples were prepared by centrifuging the blood samples at 3000 rpm for 10 minutes and was stored at 4°C. Around 0.5 ml of serum samples were taken in cryopreservation tubes for Raman spectroscopy within 36 hours of sample preparation. The Raman setup consisted of volume-phase holographic (VPH) spectrograph gratings, a high-performance deep-cooled CCD camera. The laser (single-mode diode, real-light) with a wavelength of 785 nm and 100 mW was generally used for Raman excitation. The samples were recorded 15 times for each sample, in the range of 600–1,800 cm^{-1}, with an accumulation time of 3 seconds. To check the accuracy and precision, three analysts were used to independently acquire a dataset (Yin et al. 2021).

10.5 Discussion

Zhang et al. 2021analyzed around 115 blood samples, with 76 patients in which 41 were confirmed to be infected with SARS-CoV-2 and 15 of the patients had a respiratory viral infection which was mainly caused by influenza A/B or respiratory syncytial virus (RSV) because of their flu-like symptoms identical to SARS-CoV-2. Normal samples were acquired from 20 healthy donors, and the rest of the 20 had inflammation-related diseases. ATR-FTIR analysis associated with partial least squares discriminant analysis (PLS-DA) of the individuals were able to differentiate SARS-CoV-2 infected individuals from the normal controls and other respiratory viral infections or inflammation. AUROC (area under the receiver operating characteristic curve) was 0.9561 (95% CI: 0.9071−0.9774). Serum phospholipids, along with antibodies and various serum constituents, could be depicted in the study to be significant with their distinct chemical fingerprint regions (Zhang et al. 2021).

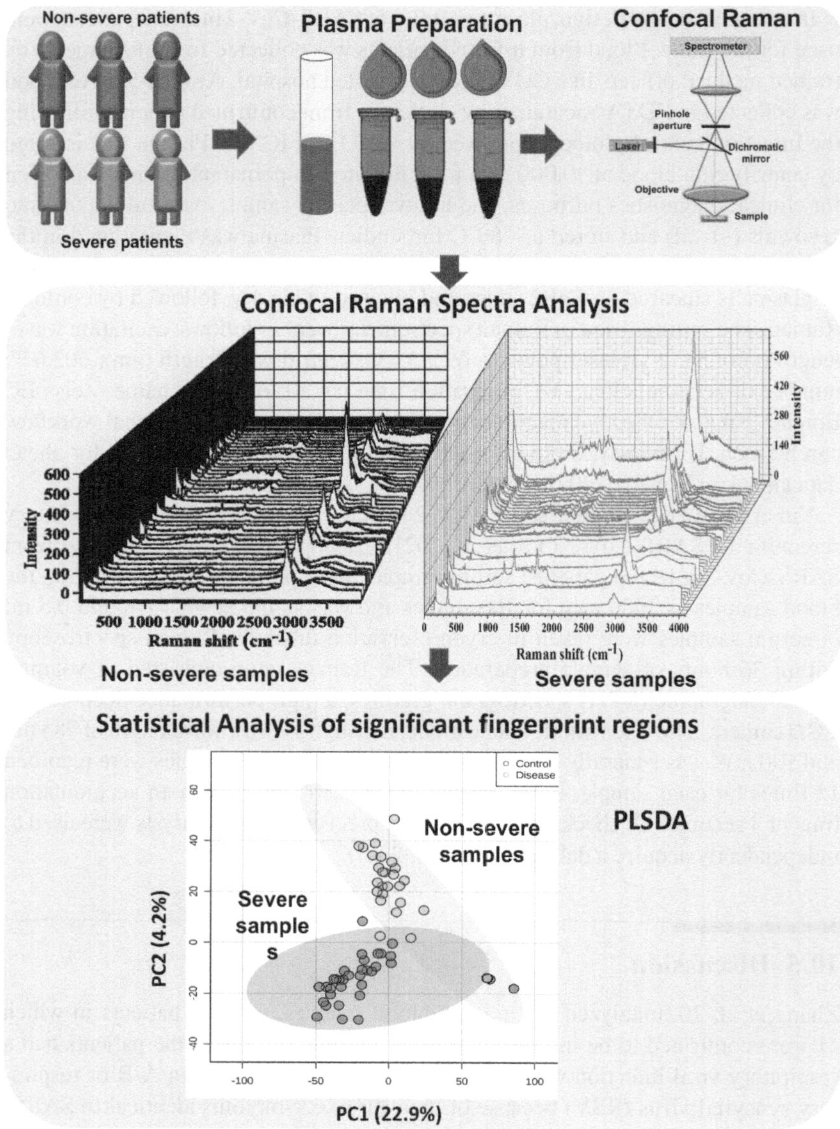

FIGURE 10.3 Schematic workflow analyzing SARS-CoV-2 plasma samples using confocal Raman spectroscopy.

Barauna et al. in 2021 analyzed 171 individuals (11 negatives and 70 positive individuals infected with SARS-CoV-2) with ATR-FTIR coupled with chemometric analysis toward screening of the SARS-CoV-2 infected patients. γ- Irradiated SARS-CoV-2 viral particles were spiked in the contrived saliva in a concentration of 1,582 copies/ml, which generated a good signal-to-noise ratio. Nucleic acid peaks along with RNA were observed predominantly in the samples. Distinct RNA fingerprint regions were also observed in the saliva samples with low copy numbers. GA-LDA (genetic algorithm-linear discriminant analysis) was used to discriminate the SARS-CoV-2 infected patients via training (50 negatives and 50 positives) and validation model (61 negatives and 20 positives) with a blinded sensitivity of 5% and specificity of 89% was achieved. Polysaccharide stretching at 1,429 cm^{-1}, asymmetric phosphate stretching due to RNA and DNA at 1,220 cm^{-1}, symmetric phosphate stretching due to nucleic acids at 1,084 and 1,041 cm^{-1}, and C–O stretching due to ribose sugar at 1,069 cm^{-1} were found to be significant chemical fingerprint regions (Barauna et al. 2021).

Yin et al. in 2021 studied a total of 177 serum samples which were categorized in three groups consisting of 63 SARS-CoV-2 infected patients, 59 suspected, and 55 healthy individuals through Raman spectroscopy coupled with machine learning support vector machine (SVM) to build the diagnostic algorithm. Further external validation was performed on 20 individuals consisting of 5 asymptomatic, 5 symptomatic patients infected with SARS-CoV-2, 5 suspected, and 5 healthy individuals. The model classified the SARS-CoV-2 infected individuals with suspected cases with an accuracy of 0.87 (95% confidence interval [CI]: 0.85–0.88) and SARS-CoV-2 infected individuals and normal individuals with an accuracy of 0.90 (95% CI: 0.89–0.91), and lastly suspected cases with normal individuals with an accuracy of 0.68 (95% CI: 0.67–0.73). SVM model classification was performed for individual test datasets (Yin et al. 2021).

10.6 Authors' perspective

We believe that due to low-noise measurements, operation complexity, and detection performance shown by FTIR-ATR and Raman spectroscopy, it can be widely used for cost-effective mass screening at hospitals, clinics, airports, and other places when coupled with multivariate analysis. In this fight to contain the pandemic, the utmost need is to detect the patients and isolate them to reduce the spread. A research group in China recently published an article where they used FTIR-ATR for rapid diagnosis of SARS-CoV-2 using serum from the patients (Zhang et al. 2021); the research group reported that the ATR-FTIR coupled with partial least squares discriminant analysis was effective to differentiate COVID-19 from normal controls and some common respiratory viral infections or inflammation, with the area under the receiver operating characteristic curve (AUROC) of 0.9561. Another research group reported a similar study in a preprint where they obtained the IR spectra using saliva samples spiked with the virus, which gave a blind sensitivity of 95% and specificity of 89% (Barauna et al. 2021). Currently, our lab at IIT Bombay is working on stratifying the SARS Cov-2 patients into risk groups so as to lower the burden on the healthcare infrastructure.

The role of Raman spectroscopy has been very significant and diverse in understanding a variety of viruses such as RNA viruses (Desai et al. 2020), porcine parvovirus (PPV), and porcine circovirus type 2 (PCV2) (Gogone et al. 2021), herpes simplex types (HSV): HSV-1, HSV-2, and varicella-zoster viruses (VZV)

(Huleihel et al. 2016), influenza A virus (Pezzotti et al. 2020), Turnip yellow mosaic virus infection (Kim et al. 2013), and hepatitis C virus (Gaggini et al. 2015). Additionally, Raman spectroscopy has contributed to discriminating bacteria and bacteriophage (Goeller and Riley 2007), dengue, and Zika non-structural protein 1 (NS1) biomarkers; typhoid and dengue fever (Naseer et al. 2019), etc. These studies are mentioned in the viral sensing section. Yet there is a necessity to incorporate artificial intelligence and discover new approaches to understand the Raman spectrum data in an efficient and variety of ways to understand disease biology in-depth.

REFERENCES

Auner, Gregory W., S. Kiran Koya, Changhe Huang, Brandy Broadbent, Micaela Trexler, Zachary Auner, Angela Elias, Katlyn Curtin Mehne, and Michelle A. Brusatori. 2018. "Applications of Raman Spectroscopy in Cancer Diagnosis." *Cancer and Metastasis Reviews* 37 (4). Springer New York LLC: 691–717. https://doi.org/10.1007/s10555-018-9770-9.

Barauna, Valério G., Maneesh N. Singh, Leonardo Leal Barbosa, Wena Dantas Marcarini, Paula Frizera Vassallo, Jose Geraldo Mill, Rodrigo Ribeiro-Rodrigues, Luciene C. G. Campos, Patrick H. Warnke, and Francis L. Martin. 2021. "Ultrarapid On-Site Detection of SARS-CoV-2 Infection Using Simple ATR-FTIR Spectroscopy and an Analysis Algorithm: High Sensitivity and Specificity." *Analytical Chemistry* 93 (5): 2950–58. https://doi.org/10.1021/acs.analchem.0c04608.

Camp, Charles H., and Marcus T. Cicerone. 2015. "Chemically Sensitive Bioimaging with Coherent Raman Scattering." *Nature Photonics*. Nature Publishing Group. https://doi.org/10.1038/nphoton.2015.60.

Desai, Sanket, Saket V. Mishra, Asim Joshi, Debashmita Sarkar, Arti Hole, Rohit Mishra, Shilpee Dutt, Murali K. Chilakapati, Sudeep Gupta, and Amit Dutt. 2020. "Raman Spectroscopy-based Detection of RNA Viruses in Saliva: A Preliminary Report." *Journal of Biophotonics* 13 (10). Wiley-VCH Verlag: e202000189. https://doi.org/10.1002/jbio.202000189.

Duncan, M. D., J. Reintjes, and T. J. Manuccia. 1982. "Scanning Coherent Anti-Stokes Raman Microscope." *Optics Letters* 7 (8): 350. https://doi.org/10.1364/ol.7.000350.

Francis, M. Mirabella. 1993. "Internal Reflection Spectroscopy. Theory and Applications." *AnalyticaChimicaActa*276(2):471.https://doi.org/10.1016/0003–2670(93)80422-h.

Gaggini, Marcio Cesar Reino, Ricardo Scarparo Navarro, Aline Reis Stefanini, Rubens Sato Sano, and Landulfo Silveira. 2015. "Correlation between METAVIR Scores and Raman Spectroscopy in Liver Lesions Induced by Hepatitis C Virus: A Preliminary Study." *Lasers in Medical Science* 30 (4): 1347–55. https://doi.org/10.1007/s10103-015-1741-6.

Gelder, Joke De, Kris De Gussem, Peter Vandenabeele, and Luc Moens. 2007. "Reference Database of Raman Spectra of Biological Molecules." *Journal of Raman Spectroscopy* 38 (9). John Wiley & Sons, Ltd: 1133–47. https://doi.org/10.1002/jrs.1734.

Giridhar, Gurram, R. R. K. N. Manepalli, and Gudimamilla Apparao. 2017. "Confocal Raman Spectroscopy."Chapter 7in *Spectroscopic Methods for Nanomaterials Characterization*, 2:141–61. Micro and Nano technologies, Elsevier. https://doi.org/10.1016/B978-0-323-46140-5.00007-8.

Goeller, Lindsay J., and Mark R. Riley. 2007. "Discrimination of Bacteria and Bacteriophages by Raman Spectroscopy and Surface-Enhanced Raman Spectroscopy." *Applied Spectroscopy* 61 (7). Appl Spectrosc: 679–85. https://doi.org/10.1366/000370207781393217.

Gogone, Izabel C. V. P., Glaucia H. Ferreira, Danielle Gava, Rejane Schaefer, Fabíola F. de Paula-Lopes, Raquel de A. Rocha, and Flavia R. O. de Barros. 2021. "Applicability of Raman Spectroscopy on Porcine Parvovirus and Porcine Circovirus Type 2 Detection." *Spectrochimica Acta—Part A: Molecular and Biomolecular Spectroscopy* 249 (March). Elsevier B.V.: 119336. https://doi.org/10.1016/j.saa.2020.119336.

Gong, Bo, Jui Hui Chen, Rieko Yajima, Yuanyuan Chen, Elaine Chase, Durga M. Chadalavada, Barbara L. Golden, Paul R. Carey, and Philip C. Bevilacqua. 2009. "Raman Crystallography of RNA." *Methods* 49 (2). Methods: 101–11. https://doi.org/10.1016/j.ymeth.2009.04.016.

Huleihel, Mahmoud, Elad Shufan, Leila Zeiri, and Ahmad Salman. 2016. "Detection of Vero Cells Infected with Herpes Simplex Types 1 and 2 and Varicella Zoster Viruses Using Raman Spectroscopy and Advanced Statistical Methods." Edited by Nancy M Sawtell. *PLoS ONE* 11 (4). Public Library of Science: e0153599. https://doi.org/10.1371/journal.pone.0153599.

Kallenbach-Thieltges, Angela, Frederik Großerüschkamp, Axel Mosig, Max Diem, Andrea Tannapfel, and Klaus Gerwert. 2013. "Immunohistochemistry, Histopathology and Infrared Spectral Histopathology of Colon Cancer Tissue Sections." *Journal of Biophotonics* 6 (1). John Wiley & Sons, Ltd: 88–100. https://doi.org/10.1002/jbio.201200132.

Khan, Rabia Sanam, and Ihtesham Ur Rehman. 2020. "Spectroscopy as a Tool for Detection and Monitoring of Coronavirus (COVID-19)." *Expert Review of Molecular Diagnostics* 20 (7). Taylor and Francis Ltd: 647–49. https://doi.org/10.1080/14737159.2020.1766968.

Kim, Saetbyeol, Sanguk Lee, Hee Youn Chi, Mi Kyeong Kim, Jeong Soo Kim, Su Heon Lee, and Hoeil Chung. 2013. "Feasibility Study for Detection of Turnip Yellow Mosaic Virus (TYMV) Infection of Chinese cabbage Plants Using Raman Spectroscopy." *Plant Pathology Journal*. Korean Society of Plant Pathology. https://doi.org/10.5423/PPJ.NT.09.2012.0147.

Lasch, Peter, Wolfgang Haensch, Dieter Naumann, and Max Diem. 2004. "Imaging of Colorectal Adenocarcinoma Using FT-IR Microspectroscopy and Cluster Analysis." *Biochimica et. Biophysica Acta—Molecular Basis of Disease* 1688 (2). Elsevier: 176–86. https://doi.org/10.1016/j.bbadis.2003.12.006.

Manoto, Sello Lebohang, Ahmed El-Hussein, Rudzani Malabi, Lebogang Thobakgale, Saturnin Ombinda-Lemboumba, Yasser A. Attia, Mohamed A. Kasem, and Patience Mthunzi-Kufa. 2021. "Exploring Optical Spectroscopic Techniques and Nanomaterials for Virus Detection." *Saudi Journal of Biological Sciences* 28 (1). Elsevier B.V.: 78–89. https://doi.org/10.1016/j.sjbs.2020.08.034.

Nabers, Andreas, Laura Perna, Julia Lange, Ute Mons, Jonas Schartner, Jörn Güldenhaupt, Kai-Uwe Saum, et al. 2018. "Amyloid Blood Biomarker Detects Alzheimer's Disease." *EMBO Molecular Medicine* 10 (5). EMBO. https://doi.org/10.15252/emmm.201708763.

Naseer, Khulla, Ayyaz Amin, Muhammad Saleem, and Javaria Qazi. 2019. "Raman Spectroscopy Based Differentiation of Typhoid and Dengue Fever in Infected Human Sera." *Spectrochimica Acta—Part A: Molecular and Biomolecular Spectroscopy* 206. Elsevier B.V.: 197–201. https://doi.org/10.1016/j.saa.2018.08.008.

Pezzotti, Giuseppe, Wenliang Zhu, Tetsuya Adachi, Satoshi Horiguchi, Elia Marin, Francesco Boschetto, Eriko Ogitani, and Osam Mazda. 2020. "Metabolic Machinery Encrypted in the Raman Spectrum of Influenza A Virus-Inoculated Mammalian Cells." *Journal of Cellular Physiology* 235 (6). Wiley-Liss Inc.: 5146–70. https://doi.org/10.1002/jcp.29392.

Pleitez, Miguel A., Tobias Lieblein, Alexander Bauer, Otto Hertzberg, Hermann Von Lilienfeld-Toal, and Werner Mäntele. 2013. "Windowless Ultrasound Photoacoustic Cell for in Vivo Mid-IR Spectroscopy of Human Epidermis: Low Interference by Changes of Air Pressure, Temperature, and Humidity Caused by Skin Contact Opens the Possibility for a Non-Invasive Monitoring of Glucose in Th." *Review of Scientific Instruments* 84 (8). Review of Scientific Instruments. https://doi.org/10.1063/1.4816723.

Salman, Ahmad, Vitaly Erukhimovitch, Marina Talyshinsky, Mahmoud Huleihil, and Mahmoud Huleihel. 2002. "FTIR Spectroscopic Method for Detection of Cells Infected with Herpes Viruses." *Biopolymers* 67 (6). John Wiley & Sons, Ltd: 406–12. https://doi.org/10.1002/bip.10171.

Yeh, Yin Ting, Kristen Gulino, Yu He Zhang, Aswathy Sabestien, Tsui Wen Chou, Bin Zhou, Zhong Lin, et al. 2020. "A Rapid and Label-Free Platform for Virus Capture and Identification from Clinical Samples." *Proceedings of the National Academy of Sciences of the United States of America* 117 (2). National Academy of Sciences: 895–901. https://doi.org/10.1073/pnas.1910113117.

Yin, Gang, Lintao Li, Shun Lu, Yu Yin, Yuanzhang Su, Yilan Zeng, Mei Luo, et al. 2021. "An Efficient Primary Screening of COVID-19 by Serum Raman Spectroscopy." *Journal of Raman Spectroscopy*, February. John Wiley and Sons Ltd, jrs.6080. https://doi.org/10.1002/jrs.6080.

Zhang, Liyang, Meng Xiao, Yao Wang, Siqi Peng, Yu Chen, Dong Zhang, Dongheyu Zhang, et al. 2021. "Fast Screening and Primary Diagnosis of COVID-19 by ATR-FT-IR." *Analytical Chemistry* 93 (4): 2191–99. https://doi.org/10.1021/acs.analchem.0c04049.

11

Role of AI and ML in empowering and solving problems linked to COVID-19 pandemic

Deeptarup Biswas, Gaurish Loya, and Graham Roy Ball

CONTENTS

List of abbreviations

AI – Artificial intelligence
COVID-19 – Coronavirus disease 2019
CT – Computed tomography

DOI: 10.1201/9781003220787-11

DL – Deep learning
Isomap – Isometric Feature Mapping
LDA – Linear Discriminant Analysis
MDS – Multidimensional Scaling
ML – Machine learning
RT-PCR – Real-time polymerase chain reaction
SARS-CoV-2 – Severe acute respiratory syndrome coronavirus 2
WHO – World Health Organization

11.1 Introduction

The COVID-19 pandemic and its associated challenges were unpredictable for the whole world, which has badly affected different professional sectors, lifestyles, and health. Serious issues like inadequate infection control guidelines, lack of transparencies and supply chain, shortage of personal protective equipment, and minimal technical support made the conditions more challenging to deal with (Palacios Cruz et al. 2021; Kumar, Rajasekharan Nayar, and Koya 2020). Artificial intelligence (AI) and machine learning (ML) played a vital role to expedite many issues linked to healthcare (Bohr and Memarzadeh 2020). AI has excellent potential for renovating the healthcare system and upgrading patient care, technology, lifestyle, social networking, data management processes, and global relationships (Hung et al. 2020; Ellahham, Ellahham, and Simsekler 2020). There has been significant investment from different companies in the worldwide healthcare industry during the last ten years. Global companies like Microsoft and IBM are developing AI- and ML-based tools to make efficient predictions in disease pathology for better and faster diagnosis (Aphinives and Aphinives 2021; Zhao et al. 2020). There has been a global initiative in upgrading robotic surgery, virtual nursing assistance, and automated administrative workflow, which could revolutionize the healthcare system (Chang et al. 2021; Gore 2020). However, the advancement of AI and ML-based strategies will take sufficient time for the complete substitution of humans.

The role of AI and ML has supported and aided in a different spectrum during this ongoing COVID-19 pandemic. Cooperation across governments, public, and private sectors has aided in the launch of many portals like the COVID-19 portal by Google (Cresswell, Williams, and Sheikh 2021). Apart from Google, multiple dashboards and web-based portals provide COVID-19 related information (Dong, Du, and Gardner 2020). The role of AI and ML has been shown in healthcare, disease management, and clinical research more. Different drug repurposing strategies based on artificial intelligence were rapidly growing, which has decreased the time taken for experimental drug discovery and vaccine development (Bohr and Memarzadeh 2020). ML-based strategies were found to be a boon during this pandemic, where scientists were able to accelerate vaccine development by predicting effective and robust binding sites (Ong et al. 2020). Imaging techniques like computed tomography (CT) were found to be an essential diagnostic aspect for severe cases, which has also been used to predict severity using AI and ML assisting diagnosis (Mei et al. 2020). Biomarker identification using omics approaches with AI and ML has also been accelerated throughout the globe, which has resulted in a better understanding of COVID-19 severity and biology (Yao et al. 2020; Shu et al. 2020). In addition, AI and ML have contributed to the classification of disease-associated issues like predicting people's mental health, country-wise risk prediction, and recovery (Dawel et al. 2020). Overall, there has been tremendous support from AI and ML, which is helping to fight COVID-19 and making life easier.

11.2 Role of big data analysis in combating the COVID-19 pandemic

The COVID-19 pandemic has shown a massive outburst of different kinds and spectrum of data: development of data lakes in industries/companies like Microsoft and Amazon, where data is stored in raw format. AWS data lake, C3.ai data lake, and Microsoft data lake were developed as one of the centralized repositories, with updated data regarding the spread and information of SARS-CoV-2 and COVID-19. These data lakes include dashboards and visualization tools, epidemiology, healthcare resources, and literature (Dagliati et al. 2021). There have been extensive developments in drug discovery and clinical trials, enhancing therapeutic strategies against COVID-19. AccessClinicalData@NIAID and ClinicalTrials.gov, both COVID-19-related studies, are two significant resources funded by NIH (National Institute of Health). A list of popular dashboards has been given in Table 11.1.

TABLE 11.1

Popular Dashboard for Tracking COVID-19

S. No.	Dashboards	Link	Summary
1	WHO COVID-19 Dashboard	https://covid19.who.int/	Spread of COVID-19 and global status of vaccine administration
2	COVID Trends	https://aatishb.com/covidtrends/	Tracks daily cases and cases of COVID-19
3	ECDC Dashboard	www.ecdc.europa.eu/en/geographical-distribution-2019-ncov-cases	Real-time updates about COVID-19 from Europe, Asia, North America, and Australia
4	John Hopkins University Dashboard	https://gisanddata.maps.arcgis.com/apps/opsdashboard/index.html#/bda7594740fd40299423467b48e9ecf6	Spread of COVID-19 around different continents of the world
5	The COVID Tracking Project	https://covidtracking.com/data/	COVID-19 positive, negative, and mortality based on testing
6	The University of Washington Dashboard	https://hgis.uw.edu/virus/	COVID-19 cases in the world and inflection points
7	University of Virginia COVID-19 Dashboard	https://nssac.bii.virginia.edu/covid-19/dashboard/	Cumulative view of infections, deaths, and recovered cases
8	WORLDOMETER	www.worldometers.info/coronavirus/	Spread of COVID-19 around different continents of the world

Genomics plays an essential role in COVID-19 research in understanding its origin and vulnerability (Mehrian-Shai et al. 2020). The application of genomics has helped identify the mutations of SARS-CoV-2, which has also helped alter the protein secondary structure (Nguyen et al. 2021). Development of databases and repository which majorly includes NCBI SARS-CoV-2 Resources (www.ncbi.nlm.nih.gov/sars-cov-2/) and other databases like CoV-GLUE, provides information on amino acid variation like replacements, insertions, and deletions found in the COVID-19 virus during this pandemic (Singer et al. 2020). IndiCoV is an extensive resource for SARS-CoV genomes and variants from India (https://clingen.igib.res.in/indicov/). The proteomics data repositories like PRIDE, PeptideAtlas, MassIVE, jPOST, approx and Panorama Public, which comes under the collaborative consortium ProteomeXchange (PX), have aided in transparent data submission and access, which has accelerated the meta-analysis and integrated omics analysis (Deutsch et al. 2020).

11.3 Role of AI and ML in drug discovery and precision medicine

The introduction of genomics in disease biology has accelerated drug discovery and precision medicine (Dugger, Platt, and Goldstein 2018). AI has been an excellent tool to achieve some of the approaches of precision medicine. Scientists are now using the thorough study of AI in risk prediction, diagnostic, drug repurposing, and development (Santus et al. 2021). Drug development strategies and precision medicine have already been implemented in cancer studies which have been possible due to the rapid development in genomics, epigenetics, and proteomics fields (Seyhan and Carini 2019). However, COVID-19 pandemics have made its utilization and implementation to higher levels. In one of the studies carried out by the University of Cambridge, where they implemented ML models to answer questions like (1) how many patients are likely to be kept on a ventilator in a coming week; (2) if there is a ventilator crisis, then which patient should be preferred based on the illness of both the patients, and (3) ICU beds count in the coming week. Answers to such problems have helped UK public hospitals to manage daily rising cases easily (Anon 2021). In research at the Tongji Hospital, Wuhan, China, by screening electronic health records of 375 patients and by implication of AI (gradient-boosted decision tree model), they have identified the patient at higher risk of mortality. They also retrieve tree prognostic features on training the model (Yan et al. 2020). These predictions-based studies along with multi-omics studies have helped in building the foundation for drug discovery research of COVID-19 using drug repurposing strategies (T. U. Singh et al. 2020). In the view of COVID-19, a study has identified 332 high-confidence proteins–human protein interactions (PPIs) experimentally and they are connected with many biological processes: 69 ligands with drugs approved by FDA along with few compounds in clinical trials (Gordon et al. 2020). One of the studies of drug repurposing in the context of COVID-19 found baricitinib, which is a potential drug used for the treatment of rheumatoid arthritis. to show potential in reducing the COVID-19 progression using modeling algorithm (Richardson et al. 2020).

COVID-19 Disease Map project is an international effort building an open-source resource for growing, arranged integration of data and knowledge bases to support pathway analysis, virus–host interaction, and various computational and disease modeling (Ostaszewski et al. 2020). In one of the studies from India, drug repurposing approach and molecular docking studies led to the identification of potential drug candidates for COVID-19 (Suvarna et al. 2021). The role of genomic analysis for an individual drug response to achieve personalized medicines and drugs is categorized in the field of pharmacogenomics (Nelson et al. 2016). An AI model like deep-docking has been developed for predicting SARS-CoV-2 epitopes to the immune system (Ton et al. 2020). A study which provided a survey on different AI-based models for drug discovery concluded that spike protein of SARS-CoV-2 is responsible for initiating the interaction of viral with the host on ACE2 receptor (Hoffmann et al. 2020).

11.4 Machine learning methods

Machine learning is a large field that superimposes and inherits ideas from various related fields and comes under artificial intelligence. ML is classified into many subcategories but especially it is divided into three major domains: supervised, unsupervised, and reinforcement learning. An overview of popular machine learning algorithms along with the classification has been shown in Figure 11.1.

11.4.1 Supervised learning

Under this learning, the machine is trained by giving a prior set of data. Any desirable model or algorithm fits the data and predicts the outcome of the new data. Supervised learning is further classified under (1) classification and (2) regression learning.

11.4.1.1 Classification

Classification, as the name suggests, helps us to classify the data into some categorical value. In the view of COVID-19, it could help predict whether or not a given sample is COVID-19 positive or negative and if it is positive, then what severity level is associated with it, by implementing ML classifier on the omic dataset. In a study, based on clinical information, blood and urine data build and compare multiple ML classifiers to classify COVID-19 severely infected patients and mild patients (Yao et al. 2020). Some of the popular classifiers are as follows:

- *Logistic regression*: Logistic output is the probability value for the given input. This method uses a logistic function in the fitting of the data. This method is employed when the output variable is of some binary nature. For example, COVID-19 Positive Group and COVID-19 Negative Group or even a binary number (0,1).

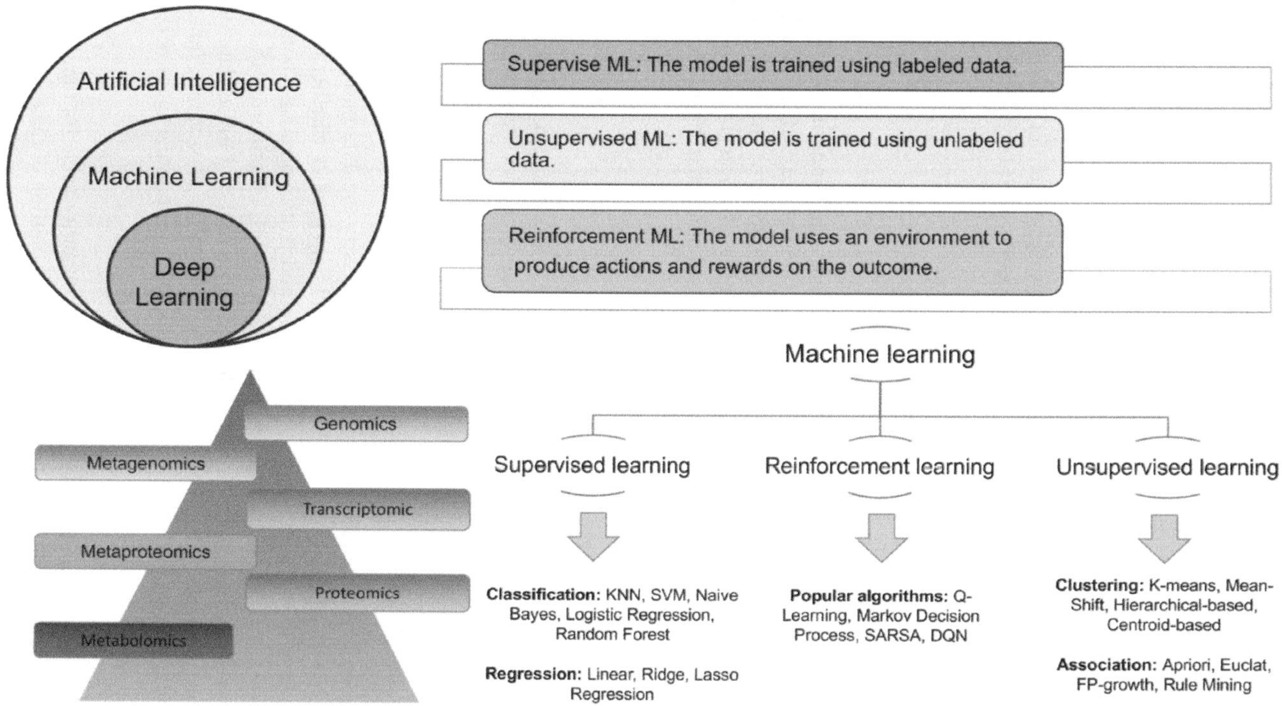

FIGURE 11.1 Overview of artificial intelligence, machine learning, and deep learning.

- *Support vector machine (SVM):* In this method, let's say we have n-dimension data. Then this n-dimension data is plotted in an n-independent axis with corresponding features value. Now, a hyperplane is chosen which best segregates the two or more classes of the variables. When new data is put into this model, then the value is predicated on how the hyperplane fits the data. This classification method has been widely used by different studies and reported as the potential strategy in plasma biomarker identification in one of the studies of COVID-19 (Suvarna et al. 2021).

11.4.1.2 Regression

Unlike classification, regression helps to predict output which has a dynamic application. In a view of COVID-19, many epidemiology studies came to notice. One work from South Korea showed the potential of popular ML models like regression, decision trees in implementing survival analysis (Muhammad et al. 2020). Some of the commonly used regression methods are as follows:

- *Linear regression:* This method is very similar to correlation. In this method, the concept of least square distance is used to fit the dataset. Once we obtain the desirable line, the prediction is made by fitting the line.
- *Ridge and Lasso regression:* These two methods are mainly used when the dataset suffers from multicollinearity, i.e., two or more independent variables are highly correlated. In such cases, it is important to reduce or remove correlated features. These models are frequently used in the classification of infection, usually when the analysis is performed on a multi-omics dataset.
- *Stepwise regression:* This method also aims for the same purposes, i.e., to reduce the features, but feature selection is done by adding/dropping covariates (if any) by forward and backward selection.

11.4.2 Unsupervised learning

Unlike supervised learning, under this learning machine is not trained by giving a prior set of data, but rather unlabeled data is given for training the model. Unsupervised learning is further classified into clustering and association learning.

11.4.2.1 Clustering

There are many different types of algorithms like k-means, centroid-based, mean-shift, and hierarchical-based. The logic was all the algorithm is to group the dataset and verify the grouping position by reiterating the process once more.

11.4.2.2 Association

Apriori, Euclat, FP-growth, and Rule Mining are some of the popular known association algorithms.

11.4.3 Reinforcement learning

This method works in the same fashion as the Markov decision method in which an agent uses an environment to produce action and reward its own output. Commonly used algorithms under this method are Q-learning, Markov decision process, SARSA, and DQN. In the pandemic time, many studies came up for the optimized lockdown period to be implemented to control the virus spread. One such study used an AI-driven approach with a reinforcement algorithm to predict the lockdown duration taking the epidemiological dataset (Khadilkar, Ganu, and Seetharam 2020).

11.4.4 Dimension reduction

Dimension reduction has become of utmost importance not only in AI/ML but also in data analysis found in many biomedical studies. Particularly, we aim to reduce our features to potential features by deleting features which are contributing to the noise in the dataset or giving an unexpected result than other dataset. The most commonly used dimension reduction algorithm used for biomedical data is principal component analysis (PCA). MetaboAnalyst is an open computational software that contains PCA analysis (Chong et al. 2018). Some of the commonly used dimension reduction algorithms are Multidimensional Scaling (MDS), Isometric Feature Mapping (Isomap), Non-linear Dimensionality Reduction Methods, and Linear Discriminant Analysis (LDA).

11.5 Role of AI and ML in the development of software, tools, and algorithms

11.5.1 MATLAB

Regression learner's app and classification app for ML and deep neural network for DL in MATLAB are elementary to use and provide great insights of data and model prediction. There are many features like descriptive statistics and visualization, probability distribution, cluster analysis, features extraction, and many ML models, which are very descriptive for explaining the significance of any biological data. This software performs all model predictions and data visualization manually and also generates code for the same (www.mathworks.com/help/stats/machine-learning-in-matlab.html).

11.5.2 Anaconda

Python is one of the most loved languages by coders. This software is very user-friendly, and by using inbuilt functions and writing very few lines of code, we can come with tremendous results. In the perspective of ML, many tools like scikit-learn provide many codes with raw data, at least for popular ML models. We all need to implement that code in our dataset. This software is encouraged to moderate to a high level of coding background (www.anaconda.com/).

11.5.3 Orange

Orange is an emerging software that has Python open-source libraries like NumPy and scikit-learn. This software unlike other programming software requires only manually placing of different widgets on the canvas interface. There are many features from data visualization, bioinformatics, to building ML models. Computational time is comparatively less than Python and it's a super user-friendly software. Even a non-programmer can easily learn and assign tasks in the orange (https://orangedatamining.com/).

11.5.4 Most popular machine learning tools

scikit-learn for Python users, shogun for C++ users: This software contains all the library and code for popular models. We need to write a similar code for our data by adjusting the data validation score and split ratio as per our discretion. Python users are encouraged to explore the scikit-learn library for ML models.

11.6 Application of machine learning in multi-omics data analysis and biomarker prediction of COVID-19

One of the recent proteome studies on the cellular processes and host factors responsible for infection was reported using the SARS-CoV-2-infected cell line model. This part has aided in the further identification and drug target and the development of therapeutic modalities (Bojkova et al. 2020). In addition, Gordon et al. have utilized affinity purification-mass spectrometry to identify viral proteins interacting with host proteins which have further been incorporated in Human Protein Atlas database (Gordon et al. 2020; Uhlen et al. 2010). Furthermore, researchers around the globe have looked at the transcriptome, proteome, and metabolome profiles of clinical samples like blood plasma, nasopharyngeal swab, and urine from COVID-19 patients to identify the underlying biological pathways that get altered due to SARS-CoV-2 infection (Shu et al. 2020; D'Alessandro et al. 2020; Shen et al. 2020). It can be safely concluded from these reports that the disease progression, morbidity, and mortality of COVID-19 is a multifactorial event caused by a diverse set of dysregulated biological pathways: hemostasis, complement cascades, leukocyte migration, regulation of cell adhesion, platelet degranulation, regulation of peptidase activity, coagulation cascades, apoptosis, T-cell signaling, neutrophil degranulation, etc.

Indian scientists and the research community has also investigated COVID-19 biology to a great extent. Institute of Genomics and Integrative Biology (IGIB) developed a paper strip–based precise and rapid diagnostic test utilizing the CRISPR/Cas system. Their method reported 100% specificity and 90.5% sensitivity as compared to RT-PCR and has the potential to augment currently available methods (P. Singh et al. 2020). A combined *in silico* and *in vitro* approach was utilized by Tiwari et al. to elucidate the mechanisms of SARS-CoV-2 pathogenesis. Another study reported deep proteome analysis of plasma and nasopharyngeal swab samples from COVID-19 patients to reveal altered pathways, biomarkers for prognosis, and drug targets (Bankar et al. 2021). The altered proteome profile in plasma samples revealed a set of significantly dysregulated pathways in severe cases of COVID-19 as compared to

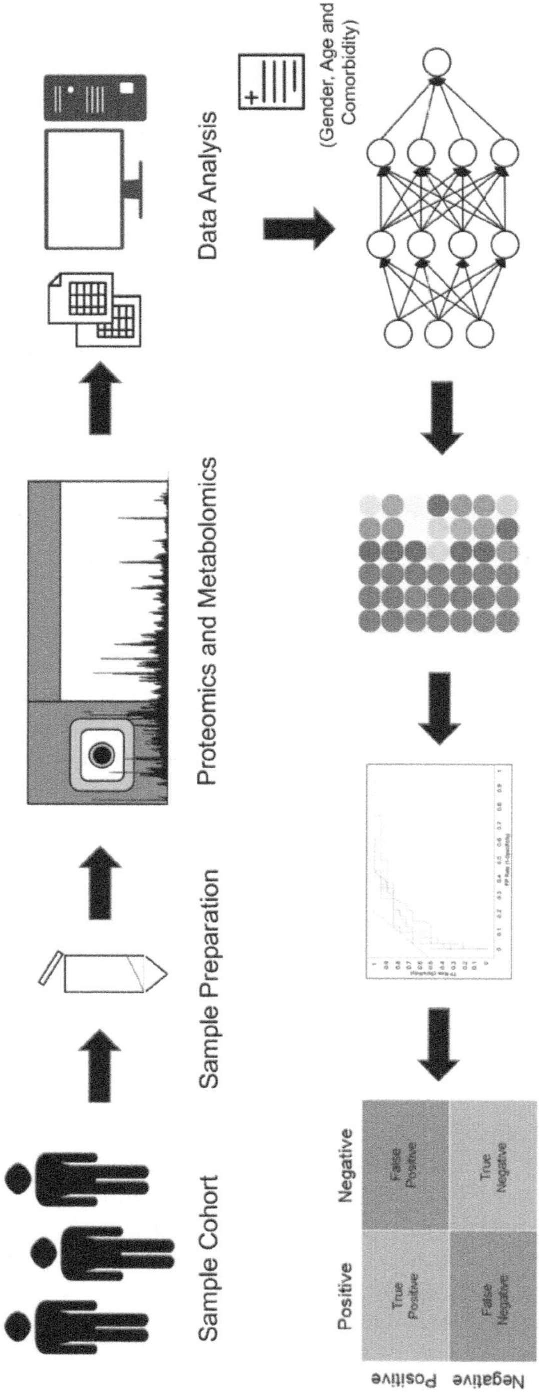

FIGURE 11.2 Schematic workflow of biomarker discovery using machine learning approaches.

non-severe ones (Suvarna et al. 2021). A supervised machine learning model helped identify prognostic markers from the plasma proteome data that could classify severe and non-severe COVID-19 patients. A tentative workflow of biomarker identification using an ML-based approach from omics datasets has been shown in Figure 11.2. Furthermore, *in silico* docking of potential drugs with host proteins involved in the identified pathways was performed for drug repurposing (Suvarna et al. 2021). A list of AI- and ML-based application in COVID-19 research has been provided in Table 11.2.

TABLE 11.2

List of Machine Learning–based Application in COVID-19 Research

S. No.	PMID	Summary of ML Approach Used in the Study	Keywords/ Dataset Used	Corresponding Author
1	33128875	Implementation of penalized logistic regression (PLR), a ML model with model training and parameter optimization gave best subset of biomarker to classify COVID-19 samples	Biomarker selection on omics dataset	Dr. Xi Zhou
2	33476063	Analysis of epidemiological data to predict early alert for spreading of communicable/ non-communicable disease Disease diagnosis by pattern recognition using medical CT images.	Epidemiological, CT scan, and X-ray radiographic data for classification of COVID-19 severity	Dr. Kuldeep Dhama
3	32568676	1020 CT (computed tomography) slices with around 108 COVID-19 positive patients and 86 patients with viral pneumonia diseases, but COVID-19 negative trained in ten well-known convolutional neural networks (CNNs) were implemented to distinguish COVID-19 positive and negative patients	CT scan data for detection of COVID-19-infected sample	Dr. Afshin Mohammadi
4	33063049	ML models like decision tree, SVM, Naive Bayes, and Random Forest were applied to the COVID-19 epidemiological dataset of patients of South Korea. Models key predictions are as follows: 1. The maximum and minimum number of days taken for given COVID-19-infected patients to recover from the infection, 2. Determining which age groups are at high risk or low risk from getting the infection	The epidemiological dataset used for survival analysis	Dr. L.J. Muhammad

(Continued)

TABLE 11.2 (Continued)

S. No.	PMID	Summary of ML Approach Used in the Study	Keywords/ Dataset Used	Corresponding Author
5	32850809	Implemented ML models like SVM, Random Forest, KNN, and Logistic Regression followed by feature selection to get 28 biomarkers for classification of severity of COVID-19 with an accuracy greater than 0.81	Biomarker selection on omics dataset	Dr. Guoqing Wang
6	33457181	Illustration of DL to Diagnostics, Protein Structure Prediction, and Drug Repurposing for COVID-19. DL has been implemented in monitoring infection spread forecasting.	Epidemiological dataset to model virus spread, other image analysis dataset	Dr. Connor Shorten
7	33046370	Took CT image of 99 COVID-19 positive patients, and used Random Forest model for classification of samples, and Regression model to predict the infection severity (moderate, severe, and critical) and recovery days, the duration of the oxygen supply needed	CT quantification for getting various patient's hospital-relevant information	Dr. Tingbo Liang
8	33539308	1. Implemented an autoML model to train and select various ML model. 2. Selected the high-accuracy model that predicted the patients' survival rate from COVID-19 infection. 3. Identified which variables (biomarkers, clinical vital signs, etc.) are dominant variables that influence model accuracy	ML model selection and survival analysis on clinical and epidemiology dataset	Dr. Morayma Reyes Gil
9	33171345	Implementation of an automatic process (with the help of 2D and 3D deep convolutional neural networks) for quantification of COVID-19 infection by extracting and selecting the image characteristics directly from the CTs and fuse them with the known clinical and biological markers	AI-based quantification on various datasets like multi-omics and CT images	Dr. Nikos Paragios

11.7 Machine learning: Empowerment during COVID-19 pandemic

The ongoing pandemic has shown a significant increase in the attention on AI and as a result a lot of development are being made in the field of AI/ML to solve or tackle the pandemic problems smartly. One such AI-driven algorithm was BluDot that deduced the SARS-CoV-2 virus in around nine days in China before WHO officially published its result (Allam, Dey, and Jones 2020). In Wuhan, a team of scientists implemented a AI robot–based algorithm which could instantly analyze the blood samples of COVID-19-infected patients to predict vital results on survival analysis (Kukar, Nayar, and Koya 2020). Deep learning model like COVID-Net emerged as an important tool which scans chest X-rays to get result if the patient is COVID-19 infectious or not (Wang, Lin, and Wong 2020). The emergence of pandemic also alerted people to a great extent and as a result, many syndromic surveillance platforms came forward which monitored and forecasted different respiratory viral outbreaks, like SARS-CoV-2 (Güemes et al. 2021). Sudden increase in the COVID cases also had an increased load in the hospital. To deal with the problem of less hospital staff and facilitating the patients in the hospital, AI robots were used in the Wuhan hospital (Dananjayan and Raj 2020).

11.8 Conclusions

The AI and ML have helped during COVID-19 pandemic and still helping in a different aspect to make life easy. With time, it has been seen that AI and ML applications have been introduced in solving COVID-19-related problems in multiple sectors like healthcare, business development, contact tracing, and data management. The biomarker identification and drug discovery have been upgraded to an advanced level after applying AI and ML, which has accelerated the diagnosis and treatment strategies during this pandemic. The prediction analysis using AI and ML has helped understand the upcoming COVID-19 waves, which might help study future pandemics.

REFERENCES

Allam, Z., G. Dey, and D. Jones. 2020. "Artificial Intelligence (AI) Provided Early Detection of the Coronavirus (COVID-19) in China and Will Influence Future Urban Health Policy Internationally." *AI* 1 (2): 156–65. www.mdpi.com/2673-2688/1/2/9.

Anon. 2021. Progress Using COVID-19 Patient Data to Train Machine Learning Models for Healthcare | University of Cambridge. Accessed April 13. www.cam.ac.uk/research/news/progress-using-covid-19-patient-data-to-train-machine-learning-models-for-healthcare.

Aphinives, C., and P. Aphinives. 2021. "Artificial Intelligence Development for Detecting Prostate Cancer in MRI." *Egyptian Journal of Radiology and Nuclear Medicine* 52 (1): 87. https://ejrnm.springeropen.com/articles/10.1186/s43055-021-00467-4.

Bankar, R., K. Suvarna, S. Ghantasala, A. Banerjee, D. Biswas, M. Choudhury, V. Palanivel, et al. 2021. "Proteomic Investigation Reveals Dominant Alterations of Neutrophil Degranulation and MRNA Translation Pathways in Patients with COVID-19." *IScience* 24 (3). https://pubmed.ncbi.nlm.nih.gov/33558857/.

Bohr, A., and K. Memarzadeh. 2020. The Rise of Artificial Intelligence in Healthcare Applications. *Artificial Intelligence in Healthcare* (January 1): 25–60.

Bojkova, D., K. Klann, B. Koch, M. Widera, D. Krause, S. Ciesek, J. Cinatl, and C. Münch. 2020. "Proteomics of SARS-CoV-2-Infected Host Cells Reveals Therapy Targets." *Nature* 583 (7816): 469–72. http://doi.org/10.1038/s41586-020-2332-7.

Chang, T. C., C. Seufert, O. Eminaga, E. Shkolyar, J. C. Hu, and J. C. Liao. 2021. Current Trends in Artificial Intelligence Application for Endourology and Robotic Surgery. *Urologic Clinics of North America*. W.B. Saunders. www.urologic.theclinics.com/article/S0094014320300690/fulltext.

Chong, J., O. Soufan, C. Li, I. Caraus, S. Li, G. Bourque, D. S. Wishart, and J. Xia. 2018. "MetaboAnalyst 4 . 0 : Towards More Transparent and Integrative Metabolomics Analysis." *Nucleic Acids Research* 46: 486–94.

Cresswell, K., R. Williams, and A. Sheikh. 2021. Using Cloud Technology in Health Care during the COVID-19 Pandemic. *The Lancet Digital Health* 3, 1 (January 1): e4–e5. http://www.thelancet.com/article/S2589750020302910/fulltext.

Dagliati, A., A. Malovini, V. Tibollo, and R. Bellazzi. 2021. "Health Informatics and EHR to Support Clinical Research in the COVID-19 Pandemic: An Overview." *Briefings in Bioinformatics* 22 (2): 812–22. https://academic.oup.com/bib/article/22/2/812/6103007.

D'Alessandro, A., T. Thomas, M. Dzieciatkowska, R. C. Hill, R. O. Francis, K. E. Hudson, J. C. Zimring, E. A. Hod, S. L. Spitalnik, and K. C. Hansen. 2020. "Serum Proteomics in COVID-19 Patients: Altered Coagulation and Complement Status as a Function of IL-6 Level." *Journal of Proteome Research*: 1–20.

Deutsch, E. W., N. Bandeira, V. Sharma, Y. Perez-riverol, J. Carver, D. J. Kundu, D. Garc, et al. 2020. "The ProteomeXchange Consortium in 2020: Enabling 'Big Data'." *Approaches in Proteomics* 48: 1145–52.

Dong, E., H. Du, and L. Gardner. 2020. "An Interactive Web-Based Dashboard to Track COVID-19 in Real Time." *The Lancet Infectious Diseases*. Lancet Publishing Group. www.who.int/china/news/.

Dananjayan, S., and G.M. Raj. 2020. Artificial Intelligence during a Pandemic: The COVID-19 Example. *The International Journal of Health Planning and Management* 35, no. 5 (September 1): 1260–1262. https://onlinelibrary.wiley.com/doi/full/10.1002/hpm.2987.

Dawel, A., Y. Shou, M. Smithson, N. Cherbuin, M. Banfield, A.L. Calear, L.M. Farrer, et al. 2020. The Effect of COVID-19 on Mental Health and Wellbeing in a Representative Sample of Australian Adults. *Frontiers in Psychiatry* 11 (October 6): 1026.

Dugger, S. A., A. Platt, and D. B. Goldstein. 2018. Drug Development in the Era of Precision Medicine. *Nature Reviews Drug Discovery*. Nature Publishing Group. www.nature.com/nrd.

Ellahham, S., N. Ellahham, and M. C. E. Simsekler. 2020. "Application of Artificial Intelligence in the Health Care Safety Context: Opportunities and Challenges." *American Journal of Medical Quality* 35 (4): 341–48. http://journals.sagepub.com/doi/10.1177/1062860619878515.

Gordon, D. E., G. M. Jang, M. Bouhaddou, J. Xu, K. Obernier, K. M. White, M. J. O'Meara, et al. 2020. "A SARS-CoV-2 Protein Interaction Map Reveals Targets for Drug Repurposing." *Nature* 583 (7816): 459–68.

Gore, J. C. 2020. Artificial Intelligence in Medical Imaging. *Magnetic Resonance Imaging*. Elsevier Inc.

Güemes, A., S. Ray, K. Aboumerhi, M.R. Desjardins, A. Kvit, A.E. Corrigan, B. Fries, et al. 2021. A Syndromic Surveillance Tool to Detect Anomalous Clusters of COVID-19 Symptoms in the United States. *Scientific Reports 2021 11:1* 11, no. 1 (February 25): 1–11. https://www.nature.com/articles/s41598-021-84145-5.

Hoffmann, M., H. Kleine-Weber, S. Schroeder, N. Krüger, T. Herrler, S. Erichsen, T. S. Schiergens, et al. 2020. "SARS-CoV-2 Cell Entry Depends on ACE2 and TMPRSS2 and Is Blocked by a Clinically Proven Protease Inhibitor." *Cell* 181 (2): 271–80.e8.

Hung, M., E. Lauren, E. S. Hon, W. C. Birmingham, J. Xu, S. Su, S. D. Hon, J. Park, P. Dang, and M. S. Lipsky. 2020. "Social Network Analysis of COVID-19 Sentiments: Application of Artificial Intelligence." *Journal of Medical Internet Research* 22 (8): e22590. www.jmir.org/2020/8/e22590.

Khadilkar, H., T. Ganu, and D. P. Seetharam. 2020. "Optimising Lockdown Policies for Epidemic Control Using Reinforcement Learning." *ArXiv*. https://doi.org/10.1007/s41403-020-00129-3.

Kumar, A., K. Rajasekharan Nayar, and S. F. Koya. 2020. "COVID-19: Challenges and Its Consequences for Rural Health Care in India." *Public Health in Practice* 1 (November): https://doi.org/100009./pmc/articles/PMC7199699/.

Mehrian-Shai, R., G. Novelli, V. Vasiliou, J. Watt, and J. K. V. Reichardt. 2020. "Genomics of COVID-19: Molecular Mechanisms Going from Susceptibility to Severity of the Disease." *Human Genomics*. BioMed Central. https://humgenomics.biomedcentral.com/articles/10.1186/s40246-020-00273-5.

Mei, X., H.C. Lee, K. yue Diao, M. Huang, B. Lin, C. Liu, Z. Xie, et al. 2020. Artificial Intelligence–Enabled Rapid Diagnosis of Patients with COVID-19. *Nature Medicine 2020 26:8* 26. 8 (May 19): 1224–1228. https://www.nature.com/articles/s41591-020-0931-3.

Muhammad, L. J., M. M. Islam, S. S. Usman, and S. I. Ayon. 2020. "Predictive Data Mining Models for Novel Coronavirus (COVID-19) Infected Patients' Recovery." *SN Computer Science* 1 (4). https://pubmed.ncbi.nlm.nih.gov/33063049/.

Nelson, M. R., T. Johnson, L. Warren, A. R. Hughes, S. L. Chissoe, C. F. Xu, and D. M. Waterworth. 2016. The Genetics of Drug Efficacy: Opportunities and Challenges. *Nature Reviews Genetics*. Nature Publishing Group. www.nature.com/nrg.

Nguyen, T. T., P. N. Pathirana, T. Nguyen, Q. V. H. Nguyen, A. Bhatti, D. C. Nguyen, D. T. Nguyen, N. D. Nguyen, D. Creighton, and M. Abdelrazek. 2021. "Genomic Mutations and Changes in Protein Secondary Structure and Solvent Accessibility of SARS-CoV-2 (COVID-19 Virus)." *Scientific Reports* 11 (1): 3487. https://doi.org/10.1038/s41598-021-83105-3.

Ostaszewski, M., A. Mazein, M. E. Gillespie, I. Kuperstein, A. Niarakis, H. Hermjakob, A. R. Pico, et al. 2020. "COVID-19 Disease Map, Building a Computational Repository of SARS-CoV-2 Virus-Host Interaction Mechanisms." *Scientific Data*. Nature Research. https://doi.org/10.1038/s41597-020-0477-8.

Ong, E., M.U. Wong, A. Huffman, and Y. He. 2020. COVID-19 Coronavirus Vaccine Design Using Reverse Vaccinology and Machine Learning. *Frontiers in Immunology* 11 (July 3): 1581. www.frontiersin.org.

Palacios Cruz, M., E. Santos, M. A. Velázquez Cervantes, and M. León Juárez. 2021. "COVID-19, a Worldwide Public Health Emergency." *Revista Clínica Española (English Edition)* 221 (1): 55–61.

Richardson, P., I. Griffin, C. Tucker, D. Smith, O. Oechsle, A. Phelan, and J. Stebbing. 2020. Baricitinib as Potential Treatment for 2019-NCoV Acute Respiratory Disease. *The Lancet*. Lancet Publishing Group. /pmc/articles/PMC7137985/.

Santus, E., N. Marino, D. Cirillo, E. Chersoni, A. Montagud, A. Santuccione Chadha, A. Valencia, K. Hughes, and C. Lindvall. 2021. "Artificial Intelligence—Aided Precision Medicine for COVID-19: Strategic Areas of Research and Development." *Journal of Medical Internet Research* 23 (3): e22453. www.jmir.org/2021/3/e22453.

Seyhan, A. A., and C. Carini. 2019. "Are Innovation and New Technologies in Precision Medicine Paving a New Era in Patients Centric Care?" *Journal of Translational Medicine*. BioMed Central. https://translational-medicine.biomedcentral.com/articles/10.1186/s12967-019-1864-9.

Shen, B., X. Yi, Y. Sun, X. Bi, J. Du, C. Zhang, S. Quan, et al. 2020. "Proteomic and Metabolomic Characterization of COVID-19 Patient Sera." *Cell*: 59–72.

Shu, T., W. Ning, D. Wu, J. Xu, Q. Han, M. Huang, X. Zou, et al. 2020. "Plasma Proteomics Identify Biomarkers and Pathogenesis of COVID-19." *Immunity* 53 (5): 1108–22.e5.

Singer, J. B., R. J. Gifford, M. Cotten, and D. L. Robertson. 2020. "CoV-GLUE: A Web Application for Tracking SARS-CoV-2 Genomic Variation." June 18. http://cov-glue.cvr.gla.ac.uk.

Singh, P., R. Chakraborty, R. Marwal, V. S. Radhakrishan, A. K. Bhaskar, H. Vashisht, M. S. Dhar, et al. 2020. "A Rapid and Sensitive Method to Detect SARS-CoV-2 Virus Using Targeted-Mass Spectrometry." *Journal of Proteins and Proteomics* 11 (3): 159–65. http://link.springer.com/10.1007/s42485-020-00044-9.

Singh, T. U., S. Parida, M. C. Lingaraju, M. Kesavan, D. Kumar, and R. K. Singh. 2020. "Drug Repurposing Approach to Fight COVID-19." *Pharmacological Reports*. Springer Science and Business Media Deutschland GmbH. /pmc/articles/PMC7474498/.

Suvarna, K., D. Biswas, M. G. Pai, A. Acharjee, R. Bankar, V. Palanivel, A. Salkar, et al. 2021. "Proteomics and Machine Learning Approaches Reveal a Set of Prognostic Markers for COVID-19 Severity with Drug Re-Purposing Potential." *Frontiers in Physiology* 12: 432.

Ton, A. T., F. Gentile, M. Hsing, F. Ban, and A. Cherkasov. 2020. "Rapid Identification of Potential Inhibitors of SARS-CoV-2 Main Protease by Deep Docking of 1.3 Billion Compounds." *Molecular Informatics* 39 (8). https://pubmed.ncbi.nlm.nih.gov/32162456/.

Uhlen, M., P. Oksvold, L. Fagerberg, E. Lundberg, K. Jonasson, M. Forsberg, M. Zwahlen, et al. 2010. "Towards a Knowledge-Based Human Protein Atlas." *Nature Biotechnology* 28 (12): 1248–50.

Wang, L., Z.Q. Lin, and A. Wong. 2020. "COVID-Net: A Tailored Deep Convolutional Neural Network Design for Detection of COVID-19 Cases from Chest X-Ray Images. *Scientific Reports 2020 10:1* 10, no. 1 (November 11): 1–12. https://www.nature.com/articles/s41598-020-76550-z.

Yan, L., H. T. Zhang, J. Goncalves, Y. Xiao, M. Wang, Y. Guo, C. Sun, et al. 2020. "A Machine Learning-Based Model for Survival Prediction in Patients with Severe COVID-19 Infection." *MedRxiv*. https://doi.org/10.1101/2020.02.27.20028027.

Yao, H., N. Zhang, R. Zhang, M. Duan, T. Xie, J. Pan, E. Peng, et al. 2020. "Severity Detection for the Coronavirus Disease 2019 (COVID-19) Patients Using a Machine Learning Model Based on the Blood and Urine Tests." *Frontiers in Cell and Developmental Biology* 8 (July 31): 683. www.frontiersin.org/article/10.3389/fcell.2020.00683/full.

Zhao, X., Y. Zhang, X. Ma, Y. Chen, J. Xi, X. Yin, H. Kang, et al. 2020. "Concordance between Treatment Recommendations Provided by IBM Watson for Oncology and a Multidisciplinary Tumor Board for Breast Cancer in China." *Japanese Journal of Clinical Oncology* 50 (8): 852–58. https://academic.oup.com/jjco/article/50/8/852/5838458.

Index

Note: Page numbers in **bold** indicate a table on the corresponding page.